Climate Change
and
Crop Growth

Climate Change *and* Crop Growth

Authors

Dr. K. Balakrishnan
Professor of Crop Physiology
Agricultural College and Research Institute
Madurai

Dr. M. Prakash
Professor, Department of Genetics and Plant Breeding
Faculty of Agriculture, Annamalai University
Chidambaram-608002

Dr. P. Jeyakumar
Professor and Head
Department of Crop Physiology
Tamil Nadu Agricultural University
Coimbatore, Tamil Nadu

SSPH

SATISH SERIAL PUBLISHING HOUSE
403, Express Tower, Commercial Complex
Azadpur, Delhi-110033 (India)
Phone : 011-27672852, Fax : 91-11-27672046
e-mail : info@satishserial.com, hkjain1975@yahoo.com
Website : www.satishserial.com

Published by :

SATISH SERIAL PUBLISHING HOUSE
403, Express Tower, Commercial Complex, Azadpur, Delhi-110033 (INDIA)
Phone : 011-27672852 Fax : 91-11-27672046
E-mail : info@satishserial.com, hkjain1975@yahoo.com

ISBN 978-93-86200-31-0 (HB)

Composed, Designed & Printed in India

Preface

The atmosphere is made up of many different gases. Several gases have the ability to reflect the sun's energy radiating from the earth back to the earth as heat. The gases with this capability are known as "greenhouse gases" (GHGs). The main GHGs are water vapour, carbon dioxide, methane, and nitrous oxide etc. Greenhouse effect is the process by which radiation from a planet's atmosphere warms the planet's surface to a temperature above what it would be without its atmosphere. Earth's natural greenhouse effect is critical to supporting life. Human activities, primarily the burning of fossil fuels and clearing of forests have intensified the greenhouse effect causing global warming. The consequent global warming may lead to very complex change in rainfall patterns and many secondary alterations in the climate system.

The changes in crop production related climatic variables will possibly have major influences on regional as well as global food production. Crop yields affected by climate change are estimated to be different in various regions, in some regions crop yields will increase, and for other regions it will decrease depending upon the latitude. To understand the various factors affecting the climate change and to relate with crop growth, it is imperative to have a look on climate and crop growth.

The book consists of 30 chapters with latest information on different aspects of climate change and crop growth. The chapters include, Green house gases, effect of temperature, CO_2 and UV. Light on physiological process, climate change on pulses, cotton and horticultural crops, carbon sequestration, nutrient acquisition and long term and short term projections on natural vegetation and ecosystem.

For writing this book, we have consulted many books, journals, research papers, review articles etc., to incorporate latest information on this topic. Nevertheless, we don't claim any originality as we have followed the generally accepted literature. The book has been written in a simple and understandable language and in a lucid manner. The book has been written primarily for the postgraduate students, teachers and researchers in Agriculture to update their knowledge on the effect of global warming and climate change on crop growth.

K. Balakrishnan
M. Prakash
P. Jeyakumar

Contents

Chapter 1

History of Climate Change

Introduction

The natural calamities like earthquakes, tsunami and volcanic eruption, though not related to weather disasters, may change the chemical composition of the atmosphere. It will, in turn, lead to weather related disasters. The largest volcano erupted at EL Salvador in 2005. It was the first of its kind and it hurled out hot rocks and ash was pumped into the atmosphere. The smoke rose more then 16 km into the atmosphere, forcing thousands of people to flee their houses. Recent studies also indicate that the Antarctic ice melting may be due to the eruption of a large volcano. The tsunami that hit the east and west coasts of India along with Indonesia on 26th December, 2004 still haunts us. Increase in aerosol (atmospheric pollutants) due to emission of greenhouse gases including black carbon and burning of fossil fuels, chlorofluorocarbons (CFCs), hydro chlorofluorocarbons (HCFCs), hydro fluorocarbons (HFCs) per fluorocarbons (PFCs), Ozone depletion and UV-B filtered radiation, eruption of volcanoes, the "human hand" in deforestation in the form of forest fires are the main causes of climate change. Paradoxically, the lack of trees also exaggerate drought in dry years by making soil to dry out more quickly. Among the greenhouse gases, CO_2 is the predominant factor leading to global warming as it traps long wave radiation and emits it back to earth surface. The global warming is nothing but heating of surface air temperature due to emission of greenhouse gases, thereby increasing global atmospheric temperature over a long period of time. Such changes in surface air temperature and rainfall over a long period of time are termed as climate change. If rainfall and temperature show year-to-year variations or cyclic trend, it is known as climate variability. Global warming is referred to as man-made interventions in emission of greenhouse gases into the atmosphere. The long-term change in rainfall or temperature, at least a definite trend for a period of 30 years, is only termed as climate change. Otherwise, interannual variations are rainfall or temperature for short periods fall under climate variability.

The global warm year – 1998

The year 1998 was the warmest one and declared the weather related year. It caused hurricane havoc in Central America and floods in China, India and Bangladesh. Canada and New England in the USA suffered heavily due to Icestorm in January 1998, while Turkey, Argentina and Paraguay with floods in June 1998. In contrast, huge crop losses were noticed in Maharashtra (India) due to unseasonal and poor distribution of rainfall during 1997-98. The 1997/1998 EL Nino event, the strongest of the last century, affected 110 million people and cost the global economy nearly US$ 100 billion. Most of the losses in recent years were recorded since 1995 onwards as the top thirteen warmest years occurred from 1995 to 2008. The estimated loss from weather related disasters in 2002 was $ 17 billion.

The global warm year-2003

The year 2003 was the year of heat and cold waves across the world. The European union suffered heavily due to heat wave that occurred in summer 2003. It was the worst heat wave in the European United since the inception of weather records. Spain, Portugal, France and Italy along with the United Kingdom suffered an economic loss $ 7 billion and the human losses accounted for 35,000.

The plant growth dropped by 30% in the European Union due to heat wave. Such heat waves are not uncommon in India during the summer season. Uttar Pradesh, Bihar, West Bengal, Orissa and Andhra Pradesh are the states that experience summer heat waves. India experienced a severe cold wave from December 2002 to January 2003. Some parts of Jammu, Punjab, Haryana, Himachal Pradesh, Bihar, Uttar Pradesh and the North Eastern States experienced the unprecedented cold wave.

Heat wave in summer 2004

The occurrence of high temperature in March 2004 adversely affected the crops like, wheat, apple, mustard, rapeseed, linseed, potato, vegetables, pea and tea across the state of Himachal Pradesh in India. The yield loss was estimated between 20% and 60% depending upon the crop. Wheat and potato harvest were advanced by 15-20 days, and the flowering of the apple was early by 15 days. The optimum temperature for fruit blossom and fruit set is 24° C in the case of apple while it experienced above 26° C for 17 days. The entire region recorded between 2.1 and 7.9° C higher maximum temperature against the normal across the state of Himachal Pradesh in March 2004 (Prasad and Rana, 2006). A decline of 39% in annual cocoa yield was noticed in 2004 when compared to that of 2003 due to rise in maximum temperature of the order of 2-3° C from 14th January in central parts of Kerala, India.

The global warm year – 2005

The year 2005 was another historic worst warm year on record for hurricanes. The hurricane Katrina over new Orleans (USA) in August, the hurricane rita in texas, central and western Cuba and southern Florida and typhoon over Hainan Province in South china and Vietnam during the week of September, while early October over Mexicans Gulf coast; heavy downpour over Mumbai on 26th July, 2005 (single- day the highest record rainfall of 944 mm) and 3rd September, 2005 over Bangalore; severe tropical storms in Andhra Pradesh during the Southwest monsoon (June- September, 2005) in India devastated cropped area to a large extent, in addition to loss of thousands of human lives. In contrast, it was declared a famine year, and food insecurity was acute in 24 sub-Sahara African countries due to drought and attack of locusts in 2005. Reports indicate that increase in sea surface temperature is likely to increase the frequency of hurricanes. Increase in its frequency during the last decade across the Pacific and Atlantic oceans was attributed to global warming, in addition to several other factors involved in the earth-atmosphere continuum.

The global warm year – 2007

The global warm year 2007 was equated to that of 1998. However, the winter 2007 was the warmest for the entire Northern Hemisphere since it began keeping records 128 years ago. It recorded 0.85°C above average of 12°C. the previous highest of 0.71°C above average, occurred in 2002 in Northern Hemisphere. Temperature during

January was over 4°C above average in Eastern Europe and Russia, more than 2.5°C above average in Canada. The entire European union recorded the warm winter, having more than 2°C above average. Untimely rains and hailstorms destroyed wheat crop of 15,000 ha over UP, Haryana and Punjab in *rabi* 2007 in India. In contrast, heavy snowfall over Kashmir Valley was recorded on 13 March, 2007 due to western disturbances.

Flood and excess rain were also noticed due to hurricanes and tropical storms worldwide in 2007. Similar was the case during monsoon 2007, causing floods across several continents (Hurricane Dean in August slammed into Mexico) including India and Bangladesh.

A huge crop loss was noticed in several states of the country due to floods in *kharif*, 2007. A major food shortage was noticed in majority of African countries due to heavy floods, which devastated several crops in region. Mali, a west African country more often plagued by droughts, received unprecedented rains during 2007.

Cold wave in winter 2008

Beijing in China had temperature as high as 16°C in February 2007, the highest since the meteorological record began in 1840, followed by one of its coldest and snowiest winter in 2008.

As a result of heavy snow since 10th January, 2008, 104 million ha of farm land was damaged and 22000 houses were destroyed. The economic loss was estimated at $ 7.5 billion.

China had lost about one-tenth of its forest resources due to severe snowstorm. A total of 17.3 million hectares of forest had been damaged across china according to the state forestry administration.

The abnormal atmospheric circulation in some regions of Europe and Asia, which had persisted for nearly twenty days since mid-January, was responsible for the impact of chilly weather, rain and snowstorm (Sharma *et al.*, 2004).

Unusual summer rains and floods - 2008

The unseasonal heavy rain were so rare in February that they occurred only the third time in the last 100 years in coastal Andhra Pradesh. The paddy crop worth Rs. 100 crore was lost due to unusual rain in March 2008 over Kuttanad and Kole lands of Thrissur district of Kerala. Similar was the case in Tamil Nadu and, to some extent, in Karnataka. On 2nd May, 2008 tropical cyclone Nargis hit the coast of Myanmar and devastated large part of the low-lying Irrawaddy delta. In rural parts of the country, up to 95 per cent of homes were wiped off and more than two million people lost their homes.

The hurricane Gustav, not a scale of hurricane Katrina, hit in September 2008 along the Louisina coastline of the United States of America and it was declared as category II with less intense. Floods in Bihar and Assam at the end of August 2008 and the first week of September, 2008 led to severe damage to the crops and many people became homeless. It was declared *national calamity* in India.

Unprecedented floods were noticed in Andhra Pradesh and Karnataka during the first week of October in 2009. All these indicate the need of proactive measures on flood management on a short-term as well as long-term basis to mitigate the ill effects of flash floods due to intense rain as projected across the world.

Rise in sea level

The mean sea level rice is likely to be slightly less than 1 mm/year along the Indian coast. However, models vary in projection of sea level rice by 2100, ranging from 28 cm to 43 cm. Sea level rise may lead to disappearance of low-lying areas of coastal belt in addition to changes in ocean biodiversity.

It threatens the health of cyclones in polar bear population. Greater number of high surges and increased occurrences of cyclones in post-monsoon period along with increased maximum wind speed are also expected as per Ministry of Environment and Forests, Govt. of India, and Department of Environment, Food and Rural Affairs (DEFRA), United Kingdom. This phenomenon of sea level rise threatens the area of land availability for farming. As per the United Nations Report of FAO, India stands to lose 125 million tonnes, equivalent to 18% of its rainfed cereal production, from climate change by 2015.

China's rainfed cereal production potential of 360 million tonnes is expected to increase by 15% during the same period. It would also cause a worldwide drop in cereal crops, put 400 million more people at risk of hunger, and put up to three billon people at risk of flooding and without access of fresh water supplies. The crop production losses due to climate change may also drastically increase the number of undernourished people, severely hindering progress in combating poverty and food security. The severest impact is likely to be in sub-Saharan African countries, which are the least able to adapt to climate change.

In 2004-2005, twenty four sub-Saharan African countries faced food emergencies, caused by a lethal combination of locusts and drought. In addition, adverse hot and dry weather in the United States and drought condition in parts of the European Union lowered cereal out put during 2005 when compared to that of 2004. The simulation models indicate that the global warming leads to reduction in rice and wheat production in northern India.

Impact of drought on foodgrain production

The Indian economy is mostly agrarian based and depends on the onset of monsoon and its further behaviour. The year 2002 is a classic example of how Indian food grains production depends on rainfall of July. It was declared All-India drought, as the rainfall deficiency was 19% against the long period average of the country and 29% of the area was affected due to drought. (The all-India drought is declared when the rainfall deficiency for the country as a whole is more than 10% of the normal and more than 20% of the country's area is affected by drought conditions). The kharif food grains production was adversely affected, registering a whopping fall of 19.1%. Similar was the situation during all-India drought in 1979, 1987 and 2009. It shows that the occurrence of drought and floods during southwest monsoon across the country affects food grain production considerably.

On regional scale also, the adverse affect on food grains production is significant due to occurrence of droughts and floods. The occurrence of floods and droughts devastated rice and other crop in Andhra Pradesh and 40% cereal production was affected in Karnataka in 2006.

It is likely to touch only a maximum of 260 million tonnes by 2020 at the present rate, through it was projected as 400 million tonnes to declare India as one of the developed countries. Hence, there is a need to augment stagnated crop productivity under the rain fed agriculture conditions through a sort of second green revolution with concerted R&D activities in the field of agriculture to feed the ever-growing population without importing food grains from outside under the projected climate change scenario. According to UNIDA, the crop production is likely to decline by 30 per cent over India by 2080 A.D (Shukla *et al.*, 2002).

Weather Abnormalities and the Global Food Price

The global food price increase is due to frequent occurrence of weather abnormalities like floods and droughts. The food grains production is not in tune with the demand whenever weather related disaster takes place. The food grain production is stagnated overall since last 10-12 years as several parts of the world experienced floods and droughts ar heat and cold waves. It was attributed to global warming. In addition, it appears that some pieces of the land cultivable under food grains are shifted to biofuels across the world over. African countries were the worst hit in 2004-05 by drought and in 2007 by flood. The crop loss varied between 40% and 90% in West African countries depending upon the severity of drought in 2004-05.

Australia and other wheat growing countries suffered heavily in 2006 due to unprecedented drought, and the wheat production was only half of the mark in 2006 as against 20-25 million tonnes of normal production in Australia.

On an average, the current price of the food items is more by 50-70% when compared to prices in 2007. The global price rice brought more than 50 million people world over below the poverty line as per the latest FAO Report.

□□□

Chapter 2

Climate Change on Monsoons

Natural swings in the climate have significantly intensified Northern Hemisphere monsoon rainfall, showing that these swings must be taken into account for climate predictions in the coming decades. Monsoon rainfall in the Northern Hemisphere impacts about 60% of the World population in Southeast Asia, West Africa and North America. Given the possible impacts of global warming, solid predictions of monsoon rainfall for the next decades are important for infrastructure planning and sustainable economic development. Such predictions, however, are very complex because they require not only pinning down how human made greenhouse gas emissions will impact the monsoons and monsoon rainfall, but also a knowledge of natural long-term climate swings, about which little is known so far. To tackle this problem, an international team of scientists around Meteorology Professor Bin Wang at the International Pacific Research Center, University of Hawaii at Manoa, examined climate data to see what happened in the Northern Hemisphere during the last three decades, a time during which the global-mean surface-air temperature rose by about 0.4°C. Current theory predicts that the Northern Hemisphere summer monsoon circulation should weaken under anthropogenic global warming.

Wang and his colleagues, however, found that over the past 30 years, the summer monsoon circulation, as well as the Hadley and Walker circulations, have all substantially intensified. This intensification has resulted in significantly greater global summer monsoon rainfall in the Northern Hemisphere than predicted from greenhouse-gas-induced warming alone: namely a 9.5% increase, compared to the anthropogenic predicted contribution of 2.6% per degree of global warming.

Most of the recent intensification is attributable to a cooling of the eastern Pacific that began in 1998. This cooling is the result of natural long-term swings in ocean surface temperatures, particularly swings in the Inter decadal Pacific Oscillation or mega-El Niño-Southern Oscillation, which has lately been in a mega-La Niña or cool phase. Another natural climate swing, called the Atlantic Multi decadal Oscillation, also contributes to the intensification of monsoon rainfall.

El Niño-Southern Oscillation

El Niño-Southern Oscillation (ENSO) is a global coupled ocean-atmosphere phenomenon. The Pacific ocean signatures, El Niño and La Niña (also written in English as El Nino and La Nina) are major temperature fluctuations in surface waters of the tropical Eastern Pacific Ocean. The names, from the Spanish for "the little boy" and "the little girl", refer to the Christ child, because the phenomenon is usually noticed around Christmas time in the Pacific Ocean off the west coast of South America. Their effect on climate in the southern hemisphere is profound. These effects were first described in 1923 by Sir Gilbert Thomas Walker from whom the Walker circulation, an important aspect of the Pacific ENSO phenomenon, takes its name. The atmospheric signature, the Southern Oscillation (SO) reflects the monthly or seasonal fluctuations in the air pressure difference between Tahiti and Darwin. El Nino affects Australia by drought.

An increase in global temperature will cause sea levels to rise and will change the amount and pattern of precipitation, and a probable expansion of subtropical deserts. Warming is expected to be strongest in the Arctic and would be associated with continuing retreat of glaciers and sea ice. Other likely effects of the warming include more frequent occurrence of extreme weather events including heat waves, droughts and heavy rainfall events, species extinctions due to shifting temperature regimes, and changes in agricultural yields. Warming and related changes will vary from region to region around the globe, with projections being more robust in some areas than others.

On the Monsoon regions of Asia (example of a tropical region)

In Bangladesh, there are three great rivers called the Ganges, the Brahmaputra and the Meghna as well as approximately 230 other rivers and tributaries. The vast majority of the country is either flat floodplain or delta areas, with highly volatile rivers reshaping the land. After heavy rains caused by the monsoon, coupled with snow melt from the Himalaya, the rivers burst their banks and there is frequent flooding. The floods have become worse and more frequent, and climate change is often blamed. In 2004, two-thirds of the country was submerged. This affected 50 million people as their homes and their staple crop, rice were submerged or washed away and millions went hungry.

It is predicted that Bangladesh will have an increase in average temperatures of 1.3°C by 2030 and 2.6°C by 2075. A higher amount of rainfall is predicted from the monsoon, increasing flooding which will destroy food, crops, housing and infrastructure. To make matters worse, a predicted near zero rainfall over the dry monsoon would be catastrophic, and cause winter drought. These extremes in climate will make life very difficult for many people living in the economically poor country. Sea level rise will also massively affect Bangladesh, swallowing upland that is already under massive pressure from erosion, population growth and land fragmentation. This will also increase salinisation of agricultural lands, further reducing agricultural productivity, salt water incursion could reach as far as 60 km inland by 2100.In Inland regions, limited drainage could result in longer periods of water logged land and lower availability of freshwater for coastal agriculture, industry and households. People could suffer immensely, from higher mortality and injuries because of flooding, greater problems with heat and freshwater stress and the potentially greater spread of infectious diseases.

Indian Monsoon in a Changing Climate

As part of the larger-scale Asian monsoon, the monsoon over India is formed due to intense solar heating in late spring as the solar maximum moves north from the equator. Northern Indian Ocean sea surface temperatures (SSTs) are warmed, along with the plains of northern India and the Tibetan Plateau. This latter region lies at an elevation of over 4500m on average, meaning that there is strong warming over a significant depth in the troposphere. The Indian Ocean south of the equator is comparatively cooler and so temperature and pressure gradients are formed from south to north. The pressure gradient leads to a cross-equatorial flow, which, when combined with the Earth's rotation, yields the familiar C-shape of boreal summer monsoon winds in the lower troposphere. These winds carry moisture evaporated from the warm

Indian Ocean to converge over the mountains on the west coast of India, before continuing to the Bay of Bengal. There they turn north and towards the west, around the low pressure "monsoon trough" of northern India, where more rain falls. The distinct difference between winter and summer wind and precipitation patterns is what characterizes the monsoon over this region (Turner and Annamalai, 2012).

Indian Monsoon Change in the Future

Simulations of future climate generally suggest an increase in monsoon rainfall on a seasonal mean, area-average basis. This is due to the twin drivers of an increasing land-sea thermal contrast, but more importantly, warming over the Indian Ocean which allows more moisture to be carried to India. Typically, increases in total rainfall over India may be in the region of 5-10%, although some climate models suggest more and some less. Climate simulations also show different patterns of rainfall change, so it is difficult to predict how rainfall might change within India.

A 5-10% change in total rainfall sounds small - does it really matter to life in India? The Indian monsoon is remarkably stable as a whole, with a mean total of around 850mm in the months of June to September, and an inter annual (year-to-year) variation of only around 10% in most cases. Even these relatively small variations in the Indian monsoon can influence things like agricultural production and the stocks and commodities market, so a 5-10% change could have significant impacts.

Floods and Droughts

For many people in India, it is the variability of rainfall on shorter time scales that has the biggest impacts - intense heavy rainfall leads to flooding; breaks in the monsoon of a week or more lead to water shortage and agricultural drought.

Floods and droughts are a normal occurrence in India. In 2002 for example, a break in the monsoon rains saw July receiving only about 50% of its normal rainfall, leading to cuts in agricultural output and declining GDP. Thinking about climate change in the context of how these extreme events will change, can help farmers and other end users to understand its implications.

It's difficult to say for certain that a particular extreme event for the monsoon is attributable to anthropogenic climate change - like the Pakistan floods of 2010. It is also thought that interannual variability of the monsoon will increase in future, whatever happens to its main driver, El Niño. Given the increasing population of the region and need for food security, improving the scientific understanding in these areas is of utmost importance.

❑❑❑

Chapter 3

Climate Change on Hydrological Cycle

Water scarcity is expected to become an ever-increasing problem in the future, for various reasons. First, the distribution of precipitation in space and time is very uneven, leading to tremendous temporal variability in water resources worldwide. For example, the Atacama Desert in Chile, the driest place on earth, receives imperceptible annual quantities of rainfall each year. On the other hand, Mawsynram, Assam, India receives over 450 inches annually. If all the freshwater on the planet were divided equally among the global population, there would be 5,000 to 6,000 m^3 of water available for everyone, every year. Second, the rate of evaporation varies a great deal, depending on temperature and relative humidity, which impacts the amount of water available to replenish groundwater supplies. The combination of shorter duration but more intense rainfall (meaning more runoff and less infiltration) combined with increased evapotranspiration (the sum of evaporation and plant transpiration from the earth's land surface to atmosphere) and increased irrigation is expected to lead to groundwater depletion.

Water

The vast majority of the Earth's water resources are salt water, with only 2.5% being fresh water. Approximately 70% of the fresh water available on the planet is frozen in the icecaps of Antarctica and Greenland leaving the remaining 30% (equal to only 0.7% of total water resources worldwide) available for consumption. From this remaining 0.7%, 87% is allocated to agricultural purposes (Kabat *et al.*, 2003). These statistics are particularly illustrative of the drastic problem of water scarcity facing the world. Water scarcity is defined as per capita supplies less than 1700 m^3/year.

According to the Comprehensive Assessment of Water Management in Agriculture, one in three people are already facing water shortages (IPCC, 2007). Around 1.2 billion people, or almost one-fifth of the world's population, live in areas of physical scarcity, while another 1.6 billion people, or almost one quarter of the world's population, live in a developing country that lacks the necessary infrastructure to take water from rivers and aquifers.

There are four main factors aggravating water scarcity according to the IPCC:

- In the last century, world population has tripled. It is expected to rise from the present 6.5 billion to 8.9 billion by 2050. Water use has been growing at more than twice the rate of population increase in the last century, and, although there is no global water scarcity as such, an increasing number of regions are chronically short of water.
- Increased urbanization will focus on the demand for water among a more concentrated population. Asian cities alone are expected to grow by 1 billion people in the next 20 years.
- As the world becomes more developed, the amount of domestic water used by each person is expected to rise significantly.
- Climate change will shrink the resources of freshwater.

The Hydrological Cycle

The hydrological cycle begins with evaporation from the surface of the ocean or land, continues as the atmosphere redistributes the water vapor to locations where it forms clouds, and then returns to the surface as precipitation. The cycle ends when the precipitation is either absorbed into the ground or runs off to the ocean, beginning the process over again.

Key changes to the hydrological cycle (associated with an increased concentration of greenhouse gases in the atmosphere and the resulting changes in climate) include:

- Changes in the seasonal distribution and amount of precipitation.
- An increase in precipitation intensity under most situations.
- Changes in the balance between snow and rain.
- Increased evapotranspiration and reduction in soil moisture.
- Changes in vegetation cover resulting from changes in temperature and precipitation.
- Consequent changes in management of land resources.
- Accelerated melting glacial ice.
- Increases in fire risk in many areas.
- Increased coastal inundation and wetland loss from sea level rise.
- Effects of CO_2 on physiological changes, leading to reduced transpiration and increased water use efficiency.

Changes in Precipitation and Drought

Projections of changes in total annual precipitation indicate that increases are likely in the tropics and at high latitudes, while decreases are likely in the sub-tropics, especially along its poleward edge. Thus, latitudinal variation is likely to affect the distribution of water resources. In general, there has been a decrease in precipitation between 10°S and 30°N since the 1980s. With the population of these sub-tropical regions increasing, water resources are likely to become more stressed in these areas, especially as climate change intensifies.

While some areas will likely experience a decrease in precipitation, others (such as the tropics and high latitudes) are expected to see increasing amounts of precipitation. More precipitation will increase a region's susceptibility to a variety of factors, including:

- Flooding
- Rate of soil erosion
- Mass movement of land
- Soil moisture availability

These factors are likely to affect key economic components of the GDP such as agricultural productivity, land values, and an area's habitability. In addition, warming accelerates the rate of surface drying, leaving less water moving in near-surface layers of soil. Less soil moisture leads to reduced downward movement of water and so less replenishment of groundwater supplies. In locations, where both precipitation and soil moisture decrease, land surface drying is magnified, and areas are left increasingly susceptible to reduced water supplies.

For each 1°C rise in temperature, global runoff will increase by 4%. Furthermore, in areas that are already vulnerable due to their limited groundwater storage availability, this cycle intensifies with increased warming and diminishing water supplies. In water stressed regions, variability of precipitation patterns is likely to further reduce groundwater recharge ability. Water availability is likely to be further exacerbated by poor management, overuse from increasing populations, and an increase in water demand primarily from increased agricultural production.

A recent global analysis of variations in the Palmer Drought Severity Index (PDSI) indicated that the area of land characterized as very dry has more than doubled since the 1970s, while the area of land characterized as very wet has slightly declined during the same time period. In certain susceptible regions, increased temperatures have already resulted in diminished water availability. Precipitations in both western Africa and southern Asia have decreased by 7.5% between 1900 and 2005.

Most of the major deserts in the world including the Namib, Kalahari, Australian, Thar, Arabian, Patagonian and North Saharan are likely to experience decreased amounts of precipitation and runoff with increased warming. In addition, both semiarid and arid areas are expected to experience a decrease and seasonal shift in flow patterns. If increased temperatures cause an intensification of the water cycle, there will be more extreme variations in weather events, as droughts will become prolonged and floods will increase in force.

Melting Glacial ice

Water supplies can also be affected by warmer winter temperatures that cause a decrease in the volume of snowpack. The result is diminished water resources during the summer months. This water supply is particularly important at the mid latitudes and in mountainous regions that depend upon glacial runoff to replenish river systems and groundwater supplies. Consequently, these areas will become increasingly susceptible to water shortages with time, because increased temperatures will initially result in a rapid rise in glacial meltwater during the summer months, followed by a decrease in melt as the size of glaciers continue to shrink. This reduction in glacial runoff water is projected to affect approximately one-sixth of the world's population.

A reduction of glacial runoff has already been observed in the Andes, whereby the usual trend of glacial replenishment during winter months has been insufficient. This is due to increased temperatures, which have caused the glaciers to retreat. It is likely that Andean communities such as El Alto in Bolivia have already observed a reduction in glacial runoff due to the scattered distribution of smaller sized glaciers, which further reduces the potential for runoff. In these areas, approximately one-third of the drinking water is dependent upon these supplies, and the recurrent trend of increased melt with diminished replenishment provides a dismal projection for water reserves if this same pattern continues.

Water Quality

Freshwater bodies have a limited capacity to process the pollution stemming from expanding urban, industrial and agricultural uses. Water quality degradation can be a major source of water scarcity. Although the IPCC projects that an increase in average temperatures of several degrees as a result of climate change will lead to an increase in average global precipitation over the course of the 21st century, this amount does not necessarily relate to an increase in the amount of potable water available. A decline in water quality can result from the increase in runoff and precipitation and while the water will carry higher levels of nutrients, it will also contain more pathogens and pollutants. These contaminants were originally stored in the groundwater reserves but the increase in precipitation will flush them out in the discharged water (IPCC, 2007).

Similarly, when drought conditions persist and groundwater reserves are depleted, the residual water that remains is often of inferior quality. This is a result of the leakage of saline or contaminated water from the land surface, the confining layers, or the adjacent water bodies that have highly concentrated quantities of contaminants. This occurs because decreased precipitation and runoff results in a concentration of pollution in the water, which leads to an increased load of microbes in waterways and drinking-water reservoirs (IPCC, 2007).

One of the most significant sources of water degradation results from an increase in water temperature. The increase in water temperatures can lead to a bloom in microbial populations, which can have a negative impact on human health. Additionally, the rise in water temperature can adversely affect different inhabitants of the ecosystem due to a species' sensitivity to temperature. The health of a body of water, such as a river, is dependent upon its ability to effectively self-purify through biodegradation, which is hindered when there is a reduced amount of dissolved oxygen. This occurs when water warms and its ability to hold oxygen decreases. Consequently, when precipitation events do occur, the contaminants are flushed into waterways and drinking reservoirs, leading to significant health implications.

Effects on Coastal Populations

For coastal populations, water quality is likely to be affected by salinization, or increased quantities of salt in water supplies. This will result from a rise in sea levels, which will increase salt concentrations in groundwater and estuaries. Sea-level rise will not only extend areas of salinity, but will also decrease freshwater availability in coastal areas. Saline intrusion is also a result of increased demand due in part to growing coastal populations that leave groundwater reserves increasingly vulnerable to contamination and diminishing water reserves.

❑❑❑

Chapter 4

Natural and Anthropogenic Activities on Greenhouse Gases Production

Earth's average temperature has risen by 1.4°F over the past century, and is projected to rise another 2 to 11.5°F over the next hundred years. Small changes in the average temperature of the planet can translate to large and potentially dangerous shifts in climate and weather. The evidence is clear. Rising global temperatures have been accompanied by changes in weather and climate. Many places have seen changes in rainfall, resulting in more floods, droughts, or intense rain, as well as more frequent and severe heat waves. The planet's oceans and glaciers have also experienced some big changes - oceans are warming and becoming more acidic, ice caps are melting, and sea levels are rising. As these and other changes become more pronounced in the coming decades, they will likely present challenges to our society and our environment.

Greenhouse Gases

Gases that trap heat in the atmosphere are called greenhouse gases. A greenhouse gas (sometimes abbreviated as **GHG**) is a gas in an atmosphere that absorbs and emits radiation within the thermal infrared range. This process is the fundamental cause of the greenhouse effect. The primary greenhouse gases in the Earth's atmosphere are water vapour, carbon dioxide, methane, nitrous oxide, and ozone. Greenhouse effect, regulated by the presence of greenhouse gases in the earth's atmosphere, is a natural process that plays a major part in shaping the earth's climate. A continued raise in GHG's has lead to enhanced greenhouse effect resulting in global warming and global climate change.

Sources of Greenhouse gases

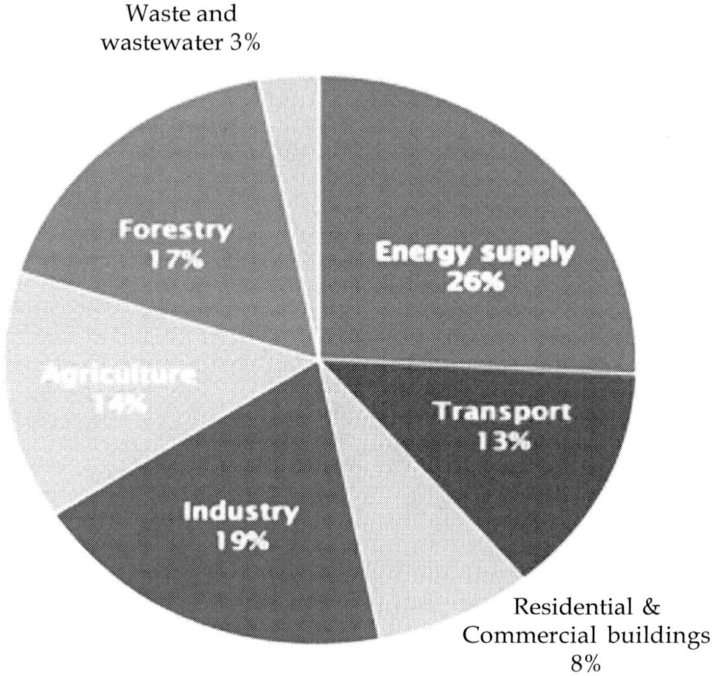

(IPCC, 2007)

Contribution of greenhouse gases to climate change

Gas	Formula	Contribution (%)
Water Vapor	H_2O	36 – 72 %
Carbon Dioxide	CO_2	9 – 26 %
Methane	CH_4	4 – 9 %
Ozone	O_3	3 – 7 %

Some of the natural and anthropogenic emissions are as follows:

Natural greenhouse gases	Anthropogenic emissions
H_2O (water vapor)	CO_2 (carbon dioxide)
CH_4 (methane)	CH_4 (methane)
CO_2 (carbon dioxide)	N_2O (nitrous oxide)
O_3 (ozone)	
N_2O (nitrous oxide)	

1. Carbon dioxide (CO_2)

Carbon dioxide enters the atmosphere through burning fossil fuels (coal, natural gas and oil), solid waste, trees and wood products, and also as a result of certain chemical reactions (e.g., manufacture of cement). Carbon dioxide is removed from the atmosphere (or "sequestered") when it is absorbed by plants as part of the biological carbon cycle.

Carbon dioxide production is not the same for every country. There is a need to reduce greenhouse gas emissions because CO_2 is a transboundary air pollutant. This basically means that pollution is not limited to the country where it is emitted. This is the same for all greenhouse gases and air pollutants.

CO_2 is the main contributor to anthropogenic radioactive forcing because of changes in concentrations from pre-industrial times; Whereas CO_2 emissions are by-and-large attributable to two major sources; energy consumption and land-use change, other emissions arise from many different sources and a large number of sectors and applications. Projections of emissions of CO_2 in 2100 range from more than 40 to less than 6 billion tonnes of elemental carbon, meaning a range from almost a sevenfold increase to roughly the same emissions level as in 1990.

2. Methane (CH_4)

Methane has a larger global warming potential than carbon dioxide. Still, emissions are lower than those of carbon dioxide. It is estimated that methane produces about one-third the amount of global warming of carbon dioxide.Methane is emitted during the production and transport of coal, natural gas, and oil. Methane emissions also result from livestock and other agricultural practices and by the decay of organic waste in municipal solid waste landfills. There are six different sources of atmospheric methane. In order of importance, these are wetlands, fossil fuels, landfills, ruminant animals, rice fields and biomass combustion.

As per IPCC projections, methane (CH_4) emissions are expected to increase most in the two regional scenarios. Increases are most pronounced in the regional wealth scenario where emissions may rise to between 549 and 1069 Mt CH_4 by 2100, compared to 310 Mt CH_4 in 1990.

3. Nitrous oxide (N_2O)

Nitrous oxide is emitted during agricultural and industrial activities, as well as during combustion of fossil fuels and solid waste. The greater part of the natural supply of nitrous oxide gas is released by oceans. Processes occurring in soils cause most of the remainder. The gas is a by-product of the biological denitrification process in anaerobic environments and of the biological nitrification process in aerobic environments. About a third of current N_2O emissions are anthropogenic. These anthropogenic emissions stem from agricultural soils, cattle feed lots and the chemical industry. N_2O emissions are generally affected by large uncertainties, because these are mainly caused by bacterial soil processes and are therefore difficult to measure.

4. Fluorinated gases

Hydrofluorocarbons, per fluorocarbons, and sulfur hexafluoride are synthetic, powerful greenhouse gases that are emitted from a variety of industrial processes. Fluorinated gases are sometimes used as substitutes for stratospheric ozone-depleting substances (e.g., chlorofluorocarbons and hydrochlorofluorocarbons). These gases are typically emitted in smaller quantities, but because they are potent greenhouse gases, they are sometimes referred to as High Global Warming Potential gases ("High GWP gases").

CFCs (chlorofluorocarbons') are gaseous compounds that contain molecules with carbon atoms bonded exclusively to fluorine or chlorine. CFCs are applied for example as refrigerants. These compounds have perhaps the greatest global warming potential among trace gases to induce global warming, because they are very persistent. CFCs absorb infrared radiation in the 8-13 μm wavelength.

Each CFC-molecule has the potential to cause the amount of global warming normally caused by tens of thousands of CO_2 molecules. The chlorofluorocarbons, $CFCl_3$ and CF_2Cl_2 have been released into the atmosphere in large quantities in the past and have long residence times. Application of CFCs is now prohibited in most countries. Many countries have agreed to reduce emissions under the 1987 Montreal protocol on substances that deplete the ozone layer.

5. Ozone (O_3)

It is present in troposphere (lower part of atmosphere). It is an important GHG resulting from industrial activities. Tropospheric O_3 is an important contributor to enhanced climate change effect.

6. Water vapor

Water vapor is a naturally occurring greenhouse gas and accounts for the largest percentage of the greenhouse gas effect between 36% and 66%. Water vapour content fluctuates regionally, but human activity does not directly affect water vapour concentrations except at local scales. Increase in atmospheric temperature caused by the greenhouse effect due to anthropogenic gases will in turn lead to an increase in water vapour content of the troposphere, with approximately constant relative humidity. The increased water vapour in turn leads to an increase in the greenhouse effect and thus a further increase in temperature; the increase in temperature leads to still further increase in atmospheric water vapour and the feedback cycle continues until equilibrium is reached. Water vapour accounts for largest percentage of GHG effect, (36 - 66% for clear sky conditions) and (66 - 85% if clouds included).

7. Small particles (Aerosols)

- Regulates the amount of energy that is absorbed and reflected by atmosphere.
- Modify properties of clouds.
- Regulates climate and to offset climate change due to increased concentration of GHGs.

The primary sources of greenhouse gas emissions are

1. CO_2

The main human activity that emits CO_2 is the combustion of fossil fuels (coal, natural gas, and oil) for energy and transportation, although certain industrial processes and land-use changes also emit CO_2. The three main sources of CO_2 emissions are described below (Gupta, 2008).

a) **Electricity:**

Electricity is a significant source of energy and is used to power homes, business, and industry. The combustion of fossil fuels to generate electricity is the largest single source of CO_2 emissions. Over 70% of our electricity comes from burning fossil fuels, mostly coal and natural gas. Type of fossil fuel used to generate electricity will emit different amounts of CO_2. Coal constituted about 90% of the total fuel mix used. To produce a given amount of electricity, burning coal will produce more CO_2 than oil or natural gas. Electricity production generates the largest share of greenhouse gas emissions. The total greenhouse gas emissions from electricity generation in 2007 was 719.31 million tonnes CO_2 eq. The CO_2 eq emissions from electricity generation were 65.4% of the total CO_2 eq emitted from the energy sector.

b) **Transportation:**

Greenhouse gas emissions from transportation primarily come from burning fossil fuel for cars, trucks, ships, trains, and planes. Over 90% of the fuel used for transportation is petroleum based, which includes gasoline and diesel. The combustion of fossil fuels such as gasoline and diesel to transport people and goods is the second largest source of CO_2 emissions. This category includes transportation sources such as highway vehicles, air travel, marine transportation, and rail.

c) **Industry:**

Greenhouse gas emissions from industry primarily come from burning fossil fuels for energy as well as greenhouse gas emissions from certain chemical reactions necessary to produce goods from raw materials. Greenhouse gas emissions from industry primarily involve fossil fuels burned on-site at facilities for energy. This sector also includes emissions from chemical, metallurgical, and mineral transformation processes not associated with energy consumption.

Many industrial processes emit CO_2 through fossil fuel combustion. Several processes also produce CO_2 emissions through chemical reactions that do not involve combustion. For example, the production and consumption of mineral products such as cement, the production of metals such as iron and steel, and the production of chemicals. It must be noted that many industrial processes also use electricity and therefore, indirectly cause the emissions from the electricity production.

2. Commercial and Residential sector

The residential sector in India is one of the largest consumers of fuel outside the energy industries. Biomass constitutes the largest portion of the total fuel mix use in this sector. Commercial and institutional sector uses oil & natural gas over and above the conventional electricity for its power needs.

Greenhouse gas emissions from businesses and homes arise primarily from fossil fuels burned for heat, the use of certain products that contain greenhouse gases and the handling of waste. Greenhouse gas emissions from this sector arise from on-site energy generation and burning fuels for heat in buildings or cooking in homes.

3. Fugitive Emissions

CH_4 escapes into the atmosphere due to mining of coal, and due to venting, flaring, transport and storage of oil and natural gas.

4. Land Use and Forestry

Land areas can act as a sink (absorbing CO_2 from the atmosphere) or a source of greenhouse gas emissions. Greenhouse gas emissions from this sector primarily include carbon dioxide (CO_2) emissions from deforestation, land clearing for agriculture, and fires or decay of peat soils. This estimate does not include the CO_2 that ecosystems remove from the atmosphere. The amount of CO_2 that is removed is subject to large uncertainty, although recent estimates indicate that on a global scale, ecosystems on land remove about twice as much CO_2 as is lost by deforestation. Global greenhouse gas emissions can also be broken down by the economic activities that lead to their production (Watson and Robert, 2000).

5. Energy Supply

The burning of coal, natural gas, and oil for electricity and heat is the largest single source of global greenhouse gas emissions.

6. Waste and Wastewater

The largest source of greenhouse gas emissions in this sector is landfill methane (CH_4), followed by methane (CH_4) and nitrous oxide (N_2O). Incineration of some waste products that were made with fossil fuels, such as plastics and synthetic textiles, also results in minor emissions of CO_2. Atmospheric concentrations of greenhouse gases that absorb infrared light lead to additional global warming, causing climate change. The impact of greenhouse gases on the enhanced greenhouse effect is determined by their residence time in the atmosphere. When atmospheric residence time is greater, the total impact of a greenhouse gas on global warming is larger. Atmospheric residence time represents the average amount of time, the molecules of a greenhouse gas exist in air before it is somehow removed. The average residence time of the greenhouse gases, carbon dioxide and nitrous oxide is more than a century.

Effect of Greenhouse Gas on Climate Change

1. Concentration or abundance

It is the amount of a particular gas in the air. Larger emissions of greenhouse gases lead to higher concentrations in the atmosphere. Greenhouse gas concentrations are measured in parts per million, parts per billion, and even parts per trillion. One part per million is equivalent to one drop of water diluted into about 60 litres of liquid.

2. How long do they stay in the atmosphere

Each of these gases can remain in the atmosphere for different amounts of time, ranging from a few years to thousands of years. All of these gases remain in the atmosphere long enough to become well mixed, meaning that the amount that is measured in the atmosphere is roughly the same all over the world, regardless of the source of the emissions.

3. How strongly do they impact global temperatures

Some gases are more effective than others at making the planet warmer and "thickening the Earth's blanket." For each greenhouse gas, a Global Warming Potential (GWP) has been calculated to reflect how long it remains in the atmosphere, on average, and how strongly it absorbs energy. Gases with a higher GWP absorb more energy, per pound, than gases with a lower GWP, and thus contribute more to warming Earth.

Chapter 5

Agricultural Activities and Greenhouse Gases Production

Introduction

Agriculture is the third largest contributor to global emissions by sector, with methane accounting for just under half of total agricultural emissions, nitrous oxide for 36 percent, and carbon dioxide for some 14 percent. Enteric fermentation or the digestion of organic materials by livestock is the largest source of agricultural emissions overall, contributing 37 percent of the total India's fertilizer industry relies heavily on natural gas (methane). Indian rice fields and livestock population are the major causes of emissions of methane and nitrous oxide.

Sources of Greenhouse gases from Agriculture

I. Agricultural Land Management

1. Rice cultivation
2. Livestock production
    - Enteric fermentation
    - Manure management
3. Agricultural soils
4. Burning of Agricultural residues in field
5. Land use, land-use change and forestry

II. Agricultural Inputs and Farm Machinery

1. Fuel and Electricity
2. Fertilizers and agricultural lime
3. Pesticides
4. Irrigation
5. Harvesting and threshing
6. Farm machinery

I. Agricultural Land Management

Increased growth in agricultural production has also resulted in increased agricultural greenhouse gas emissions with a huge proportion of emissions coming from livestock production.

1. Rice cultivation

The acreage under rice cultivation in India is estimated at about 44 million hectares, mostly planted in wet monsoon or irrigated systems by flooding and puddling fields. These rice fields are a major source of GHGs like CH_4 and N_2O. Researchers have attempted to estimate GHG in rice fields. But there are uncertainities in Indian rice fields due to diverse soil and climatic conditions and crop management practices. Carbon dioxide emissions from cultivated organic soils account for some 14 percent of total agricultural emissions, with Asia contributing 54 percent of these emissions.

Wetland rice

Rice lands can be classified as irrigated, rain-fed, deepwater and upland. In irrigated rice fields, the floodwater is fully controlled and kept shallow. In rain-fed rice fields, precipitation controls flooding of soils. At times in the growing season, soils of rain-fed rice fields may dry up or be flooded up to 50 cm. In deepwater rice fields, floodwater rises to more than 50 cm during the growing season, and it may reach several meters. Upland rice fields are neither flooded nor does the topsoil become water saturated at any significant period of time. Most rice is grown on wetlands; only 13% is cultivated on uplands as is wheat or maize.

Wetland rice soils

Wetland rice soils have free water at the surface for at least the major part of the crop growing season. Flooding drastically reduces the diffusion of atmospheric oxygen into the soil, and facultative and anaerobic microorganisms sequentially reduce soil substrates. The redox potential is a quantitative indicator measuring the tendencies of different oxidations and reductions (net flow of electrons) to occur. The higher the value of the redox potential, the greater the presence of strong oxidizing agents in a soil.

Methane formation

Methane is produced as the terminal step of the anaerobic breakdown of organic matter in wetland rice soils. Methane is exclusively produced by methanogenic bacteria that can metabolize only in the strict absence of free oxygen and at redox potentials of less than -150 mV. Most methanogens are neutrophilic, with an optimal pH of 6-8. Methanogens rely on a plethora of other microorganisms to provide them with the few substrates they can catabolize: hydrogen, carbon dioxide, formate, acetate, methanol, methylamines, and methyl sulphids. In wetland rice soils, methane is largely produced by transmethylation of acetic acid and, to some extent, by the reduction of carbon dioxide. The formation of methane is preceded by the production of volatile acids.

Methane oxidation

Methane-oxidizing bacteria (methanotrophs) are abundant in the oxidized floodwater-soil interface and in the rice rhizosphere. They sequentially oxidize methane to carbon dioxide via methanol, formaldehyde, and formate. Oxygen is essential for the growth of methanotrophs, but the required partial pressure may be low. Methane oxidation greatly limits diffusion of methane to the atmosphere. Up to 60% of the methane produced during a rice growing season may be oxidized before it reaches the atmosphere.

Rice plants supply atmospheric oxygen to the roots for respiration via a special vascular system, the aerenchyma. The aerenchyma has its own openings at the leaf sheath, and the gas supply to and from the roots is independent of transpiration and stomatal gas exchange. Oxygen diffusion from rice roots constitutes an important part of the roots' oxidizing power, aside from enzymatic hydrogen peroxide production. Because of the abundance of methane-oxidizing bacteria present in the rhizosphere, the rhizophere's potential for methane oxidation is high. Ten times more methane-oxidizing bacteria in the rhizosphere was counted than in the bulk anaerobic soil and one-third more than in the oxidized soil-water interface.

2. Livestock production

India has the largest cattle, buffalo and goat population and is fifth in sheep population in the world. Cattle and small ruminants are the most dominant features of the agrarian economy and also a major source of GHG emission. The greenhouse gas emission from livestock has three components

- CH_4 emission from enteric fermentation
- Manure management
- N_2O emission from animal wastes

Livestock contribute to global emissions in other ways as well. Manure deposited and left on pastures is a major source of nitrous oxide emissions because of its high nitrogen content. When more nitrogen is added to soil than is needed, bacteria convert the extra nitrogen into nitrous oxide and release it into the atmosphere. Emissions from manure on pasture in Asia, Africa, and South America together account for as much as 81 percent of global emissions from this source. These emissions from the three regions increased 42 percent on average between 1990 and 2010, reflecting an increase in range-based livestock populations; elsewhere, these emissions either decreased or stagnated.

2.1. Enteric fermentation

Methane is produced as part of the normal digestive processes in animals. During digestion, microbes in the animal's digestive system ferment feed. This process called enteric fermentation, produces methane as a by-product. Enteric fermentation in the digestive system of ruminants leads to production and emission of CH_4. These animals possess rumen, which allows them to digest cellulose and other roughages found in plant material. A small fraction of symbiotic microorganisms (3-10%) is methanogenic bacteria which produce methane while removing methane from the rumen. In ruminant animals principally cattle, buffalos, sheep and goat, a significant amount of fermentation takes place in the rumen resulting in relatively large CH_4 emissions per unit of feed energy consumed. Beef cattle account for about 70 percent and dairy cattle for about 25 percent of these methane emissions. If beef and dairy cattle numbers increase, methane emissions will also increase. Feed qualify and feed intake influence the level of methane emissions. In general, lower feed quality and higher feed intake lead to higher methane emissions.

2.2. Manure management

Management decisions about manure disposal and storage affect emissions of CH_4 and N_2O, which are formed in decomposing manures as a byproduct of methanogenesis and nitrification/ denitrification. Livestock manure is principally composed of organic material. When this organic material decomposes under anaerobic environment, methanogenic bacteria produces methane. Methane is produced by the anaerobic (without oxygen) decomposition of manure. When the manure is stored or treated as liquid, it tends to decompose anaerobically and produce a significant quantity of methane. When manure is handled as solid or deposited on pastures, it tend to decompose aerobically and a little or no methane is produced. Dairy cattle and swine produce about 85 percent of the methane emissions. Methane emissions will increase as the number of large scale livestock confinement systems increases.

Methane emissions can be reduced through the application of technologies designed to capture the methane and use it as an energy source. In addition to reducing methane emissions, methane capture will improve the profitability of the livestock operation by offsetting the need for fossil fuel energy from outside sources.

Three potential sources of N_2O emissions related to livestock production (Swamy and Bhattacharya, 2011) are :

1. Animal themselves
2. Animal wastes during storage and treatment
3. Dung and urine deposited by free-range grazing animals.

Livestock manure is a valuable resource for nutrients and an excellent soil amendment to improve soil quality, tilth and productivity. The main GHGs emitted by manure are methane (CH_4), which is emitted during the anaerobic (without oxygen) decomposition of organic matter during storage and nitrous oxide (N_2O), which is emitted during storage and soil application. Additional gases emitted from manure include: ammonia (NH_3) and nitrogen oxides (NO_X), which contributes to odour and are indirect sources of nitrous oxide.

Manure management is an essential practice in minimizing GHG emissions caused by microbial activities during manure decomposition. Factors that affect GHG emissions from manure include temperature, oxygen level (aeration), moisture, and sources of nutrients. These factors are affected, in turn, by manure type (livestock type), diet, storage and handling of manure (pile, anaerobic lagoon, etc), and manure application (injected, incorporated, etc). It is important to note that proper manure management is essential for any agricultural operation because improper use of manure can lead to negative impacts on the environment.

3. Agricultural soils

Direct and indirect emissions from agricultural soil are determined by a multitude of factors such as rate of fertilizer and organic manure application, yield and area under cultivation. Direct emission sources include N fertilizers, crop residues and mineralization process of soil organic matter. Indirect sources comprise leaching, run-off and atmospheric deposition. Nitrous oxide is produced naturally in soils through the microbial processes of nitrification and de-nitrification. During nitrification, ammonium (NH_4^+) is reduced to nitrates (NO_3^-). During de-nitrification, nitrates (NO_3^-) are reduced to nitrogen gas (N_2). An intermediate step in both of these processes is the creation of nitrous oxide (N_2O). Both the process are influenced and controlled by environmental conditions. They are independent of the origin of N, whether organic or mineral fertilizers or from symbiotic N fixation in legumes.

The large increase in the use of nitrogen fertilizer for the production of high nitrogen consuming crops like maize has increased the emissions of nitrous oxide. Although nitrogen fertilizer is essential for profitable crop production, the development of practices for more efficiently using nitrogen fertilizer has the potential to significantly reduce nitrous oxide emissions while also reducing production costs and mitigating the nitrogen contamination of surface and ground waters.

4. Burning of agricultural residues in field

Agricultural wastes are burnt in the field to clear remaining straw and stubble after harvest and to prepare the field for the next cropping cycle. Farmers prefer crop residues burning as a quick and labour saving process to dispose of the crop residues of rice,wheat, maize and sugarcane. Emission of CO_2 during burning of crop residues are considered neutral, as it is reabsorbed during the next growing season. Biomass burning is one of the significant source of atmospheric aerosols and trace gas emissions which has a major impact on human health. In addition to aerosol particles, biomass burning due to forest fires and crop residue burning are considered a major source of carbon dioxide, carbon monoxide, methane, volatile organic compounds, nitrogen oxides and halogen compounds.

5. Land use, land-use change and forestry

Plants absorb carbon dioxide (CO_2) from the atmosphere as they grow, and they store some of the carbon throughout their lifetime. Soils can also store CO_2, depending on how the soil is managed. This storage of carbon in plants and soils is called biological carbon sequestration. Because biological sequestration takes CO_2 out of the atmosphere, it is also called a greenhouse gas "sink."

Emissions or sequestration of CO_2 can occur as land uses change. For example, CO_2 is exchanged between the atmosphere and the plants and soils on land as former cropland is converted into grassland, as new areas are cultivated and become cropland, or as forests grow. In addition, using biological feedstocks (such as energy crops or wood) for purposes such as electricity generation, input to processes that create liquid fuels, or building materials can lead to emissions or sequestration.

II. Agricultural inputs and Farm Machinery

1. Fuel and Electricity

Use of fossil fuels in agriculture results in CO_2 emissions and there are additional emissions associated with production and delivery of fuels to the farm. Carbon emissions attributed to fossil fuels are estimated using existing C coefficients, higher heating values, fuel chemistry and the energy consumed during production and transport of fuels. The CO_2 emission attributed to electricity consumption are based on the fuels used in power generation.

2. Fertilizer and agricultural lime

The production of fertilizers demands much energy and generates considerable GHG emissions. Fertilizer production consumes approximately 1.2% of world's energy and is responsible for approximately 1.2% of the total GHG emissions. The fertilizer industry deals primarily with supplying N,P and K although chemical fertilizers are used to supply 13 essential plant nutrients. This analysis includes the three primary nutrients and agricultural lime in the form of crushed limestone ($CaCO_3$).

3. Pesticides

Modern pesticides are almost entirely produced from crude petroleum or natural gas products. The total energy input is thus both the material used as feed stock and the direct energy inputs. Carbon dioxide emissions from production of pesticides consists of both these contributions to manufacture the active ingredient. Post production emissions include those from formulation of the active ingredients into emusifiable oils, wettable powders or granules and those from packaging, transportation and application of the pesticide formulation. Carbon dioxide emissions from pesticides use are estimated for specific classes by calculating average values of energy input for the production and application of individual pesticides.

4. Irrigation

The on-farm wells, on-farm surface reservoirs and off-farm surface reservoirs are the major sources of irrigation water. Fossil fuel used to power pumps which distribute irrigation water were calculated using energy expenses for on-farm pumping and energy price estimates. The energy use and C emissions from pumping water were applied to both on-farm and off-farm surface reservoirs.

5. Harvesting and threshing

Energy and CO_2 emissions during harvesting and threshing of agricultural produce is also important. Greater the productivity, greater is energy required and emissions for harvesting and threshing.

6. Farm machinery

Energy and CO_2 emissions associated with different tillage practices are a consequence of the fuel used by farm machines and the energy consumed in manufacture, transportation and repair of the machines. While CO_2 emissions associated with the application of fertilizers and pesticides were calculated along with other farm operations, they do not occur on all fields and in all years as do other farm operations. Therefore, CO_2 emissions from the application of fertilizers and pesticides are weighted by their extent of application (Table 1).

Table 1. Methods to reduce GHG from Agriculture Sector

Type	How Emissions are Reduced	Examples
Land and Crop Management	Adjusting the methods for managing land and growing crops.	Fertilizing crops with the precise amount of nitrogen required, since less efficient nitrogen application can lead to higher N_2O emissions.Draining water from wetland rice soils during the growing season to reduce CH_4 emissions.
Livestock Management	Adjusting feeding practices and other management methods to reduce the amount of CH_4 resulting from enteric fermentation.	Improving pasture quality to increase animal productivity, which can reduce the amount of CH_4 emitted per unit of animal product. Also, increased productivity can be accomplished through breeding.
Manure Management	Controlling the way in which manure decomposes to reduce N_2O and CH_4 emissions.Capturing CH_4 from manure decomposition to produce renewable energy.	Handling manure as a solid or depositing it on pasture rather than storing it in a liquid-based system such as a lagoon. This would likely reduce CH_4 emissions but may increase N_2O emissions. Storing manure in anaerobic containment areas to maximize CH_4 production and then capturing the CH_4 to use as an energy substitute for fossil fuels.

Climate change is the most significant environmental concern of our lifetimes, and comprehensive action is essential. Agriculture and food sector are the important factors to climate change. Approximately one third of GHG's emissions are produced by agriculture and land use changes, with 18% of the overall total coming from livestock alone. Food security for the growing population is positioned to remain a challenge in the next few decades. But from the environmental perspective, greater agricultural activity also increase emissions of GHG's which contribute to climate change. The question facing the agricultural production is how to produce the required quantity of food with the lowest possible GHG emissions. Future research should focus on management of natural resources, particularly soil and water, to sustain a long-term productivity with minimum environmental pollution.

❑❑❑

Chapter 6

Carbon Dioxide on Crop Physiological Processes

Plants directly influenced by rising CO_2 and it is a first molecular link between atmosphere and biosphere. Affect from microscopic cellular organisms to macroscopic agro-ecosystem. Potentially influence the photosynthesis, respiration and transpiration processes

A) Single leaf response to carbon dioxide

1. Leaf Carbon Dynamics

Short-term impact of elevated CO_2 provide good indicator of photosynthetic response of single leaves Long-term leaf response to elevated CO_2 altered by abiotic variables such as N availability or changes in PAR or temperatures. Eg. High temperature favours PCO cycle in C_3; but, PCO suppression by elevated CO_2 stimulates photosynthesis as temperature increases

2. Inhibition of Dark Respiration

During the oxidational of organic compound, CO_2 is released in mitochondria. Very high CO_2 concentration (1000 ppm) drastically reduces dark respiration rates. Inhibition of cytochrome oxidase and succinate dehydrogenase occur with elevated carbondioxide. Elevated CO_2 reduces leaf dark respiration rate because of poor translocation and low NO_3 reduction during night.

3. Leaf biochemistry

Elevated CO_2 increases C/N ratio in leaf leads to increase in CHOs, lignin and/or cellulose contents. Changes in decomposition rates, reductions in N cycles and changes in freezing-resistance occur. Carbon based secondary chemical production favours plant-herbivore relations. Herbivore feeding strongly influenced by leaf allelochemicals as well as leaf nutritional quality in under elevated CO_2 level

4. Response of stomatal activities to elevated carbon dioxide

Stomatal conductance decreased under elevated CO_2 changes water use by crops (more WUE). High CO_2 inhibits the stomatal activities lead to reduction of photosynthesis. High WUE due to increase or maintainence of photosynthesis mostly occur in C_3 crops.

B) Whole plant response to elevated carbon dioxide

1. Plant Development

- Stimulation in plant growth
- Enhanced seed germination with the enhancement of ethylene production
- Vegetative development particularly sensitive as CO_2 increases tiller production in both C_3 and C_4 crops
- Stimulates leaf production, growth and size of woody C_3 species in water-stressed environment
- Improved WUE & leaf water content under elevated CO_2 is a significant factor for higher photosynthetic rates in C_4 plants under salinity or water stress conditions
- Promotion of root growth due to increased root length, root diameter and cortex width.
- Increase of fine-root colonization with VAM myccorhizae and increase of root nodules.
- Increases floral number and pollen production, seed and fruit size, number & quality.
- Reduction of seed N content in non-leguminous crops.
- Asexual production showed increased response.
- Shifts in need based photosynthate allocations among organs.
- On setting of reproductive phase altered by faster growth rates.
- Altered plant senescence - advancing in some crops and delaying in others

2. Carbon Dynamics

- Stimulation of photosynthesis / acclimation function of sources and sinks of carbon.
- Elevated CO_2 induces temporal (time related phases) changes in plant development, thus alters sink availability.
- In sugarbeet, leaf development altered due to additional carbon sinks.
- Self-shading and N redistribution within whole canopy alters photosynthetic adjustment in leaves.
- In rice, N limits tiller production under elevated $CO_{2,}$ cause subsequent decline in photosynthesis.

3. Stomatal Regulation and Water Use

- For whole plant and communities, aerodynamic resistance(force which prevents the movement of a liquid or gas) is important in controlling the the stomatal activities

4. Plant Competition

- Crop-weed competition vary greatly in different regions due to variations in temperature, precipitation, soil characters etc.
- C_3 and C_4 crops interact with C_3 and C_4 weeds
- Competition among crops and weeds under elevated CO_2 remain meagre.
- Competition of C_3 crop with a C_4 weed under elevated CO_2, increased crop/weed biomass ratio.
- Biomass/grain yield of C_4 crop reduced when grown with C_3 weeds.

5. Elevated temperature and CO_2

Elevated CO_2 is expected to enhance leaf level photosynthetic rates by 30-40% in plants with C_3 metabolism, like soybean, rice, and wheat. Much less effect on leaf photosynthesis will be observed in plants with C_4 metabolism because the C_4 enzyme system increases the concentration of CO_2 at the site of photosynthesis. Plants that utilize C_4 metabolism will be less responsive to increases in CO_2, because the enzyme system is nearly saturated at present-day atmospheric CO_2. The photosynthetic process is directly affected by the interaction of elevated temperature and CO_2 and the biochemical aspects of photosynthesis have been utilized to create mechanistic plant growth models. The model equations have been repeatedly validated with field observations and then used to predict consequences of climate change on plants. Thus, the optimum temperature for C_3 photosynthesis is projected to increase with elevated CO_2. Dark respiration is associated with biosynthesis of new tissue and maintenance of old tissue. As temperature increases, the rate of dark respiration increases, thereby using more carbohydrates for growth and maintenance. Elevated CO_2 has no effect on mitochondrial respiration.

Changes in Physiological Processes

1. Respiration

- Both short- and long term exposure of crops to elevated CO_2, reduce respiration.
- Growth and maintenance respiration will be affected due to change in leaf N concentration.

Photorespiration is directly associated with the principle photosynthetic enzyme, rubisco (ribulose biphosphate carboxylase oxygenase). Rubisco functions in either of two ways: either it will fix CO_2 in photosynthesis (carboxylase activity) or it will fix oxygen (oxygenase activity), which leads to photorespiration. Elevated temperature will increase the extent of oxygenase activity because the solubility of O_2 increases in relation to CO_2 as temperature increases, as does the affinity of rubisco for O_2 relative to CO_2. Plants with C_3 metabolism produce glycolate as a product of oxygenation, providing a substrate for photorespiration. Elevated atmospheric CO_2 will decrease photorespiration at high temperatures by increasing the amount of substrate for carboxylation. Mostly tropical C_4 plants have an additional enzyme (PEP carboxylase) which concentrates CO_2 at the site of photosynthesis, and avoid photorespiration because the concentrated CO_2 effectively competes against O_2 for rubisco-binding sites.

2. Transpiration

- Elevated CO_2 reduces transpiration by reducing stomatal conductance in most species.

- Reduction of stomatal conductance about 40% (due to high CO_2), caused 10 % reduction in canopy water requirement or loss by transpiration.
- In rice, stomata of lower epidermis are more sensitive to elevated CO_2 attributes to reduction in stomatal conductance. Increase in Meta- and proto-xylem favours more transfer of water and nutrients from soil to plants increases plant water status.

3. Stomatal Conductance

Stomata are tiny openings or valves distributed on leaf surfaces that allow for gas exchange. Many studies have shown a direct effect of CO_2 and temperature on stomatal conductance. As temperature increases, stomatal apertures open to allow evaporative cooling through the release of water vapor. When stomatal pores open, the supply of CO_2 available for the photosynthetic enzymes is increased. In contrast, stomatal density and conductance decrease when plants are grown with elevated CO_2. For example, stomatal conductance decreases about 30–40% in many plant species when grown under a double concentration of CO_2. Since stomata provide the entry for CO_2 molecules into leaves, plants have developed the ability to sense higher CO_2 concentration and regulate the amount of CO_2 by closing the stomata. The degree of stomatal opening is decreased at elevated CO_2, thereby decreasing water loss through transpiration and also leading to a small increase in leaf temperature. Thus, just as elevated atmospheric CO_2 leads to increased atmospheric temperatures, the temperature of individual plant leaves and cells will increase. Since a steady increase in atmospheric CO_2 has been measured, the direct increase in leaf temperature due to increased $CO_2$and the effect of increased temperature on physiological processes should be recognized as an important concern. Since leaf cooling through evapo-transpiration is another function of these openings, decreased stomatal conductance due to elevated CO_2 may contribute to increased leaf temperature.

4. Nitrogen Assimilation

Lesser RUBISCO and photosynthetic apparatus due to reduction in photosynthetic activities, reduces leaf N content.

- Elevated structural and non-structural CHO's increases the C/N ratio and decreases the protein content both in leaves as well as in seeds.

- Shorter exposure to elevated CO_2 diverts the role of reductants for the reduction of NO_3^- or NO_2^- reduction to photosynthetic CO_2 fixation.

5. Water use efficiency (WUE)

- Under water stress, plants dry slowly under elevated CO_2 level due to lower **stomatal conductance** and lower **evapo-transpiration**.
- Elevated CO_2 increases drought tolerance by lowering osmotic potential and maintains turgidity; growth is continued under water stress.
- Plants under elevated CO_2 utilize less water and increase the water use efficiency.
- Depletion of soil water occurs at lower rate.
- Water-stressed plants maintain high RWC under elevated CO_2.
- Conservation of soil water than osmotic adjustment accounting for greater plant water content
- Increase in WUE under elevated CO_2 to the level of 33%
- Increased WUE due to increase in photosynthesis and reduced water loss due to partial closure of stomata.

❑❑❑

Chapter 7

Temperature on Crop Physiological Processes

Every plant has an optimum temperature and season for growth, development, and reproduction. Increased temperature will affect ecosystems and crops differently according to region. With warming, slight shifts in regional production of food crops will occur. For example, some plants require a chilling period to break dormancy and/or to produce flavorful fruit. Negative impacts are likely in areas where high temperature or less than adequate rain already limits productivity. Beneficial results are expected in areas where cold temperatures currently limit crop yield. Unfortunately, high temperature stress is among the least understood of all plant processes.

Increase of GHGs increases global mean temperature. Global mean temperature likely to be 1°C above than present value by 2025 AD and by 3 to 4.5°C above by 2100 AD. The mean temperature across india has risen by 0.5°C during the period 1960 and 2009. In temperate latitude, rise of temperature increase photosynthesis and respiration of plants and microbes. At particular high temperature, respiration increases more than photosynthesis. Global warming likely to increase CO_2 concentration of atmosphere, acts as feed back to global warming. Temperature increase due to high CO_2 become autocatalytic component to global warming.

Mechanism of Heat Tolerance

- Crops adapt to prevalent soil and weather conditions.
- Except for transpirational cooling, plants unable to adjust their tissues to temperature significantly.

- Crops have several mechanisms to tolerate higher temperature stress.
- Adaptation mechanisms to high temperature reflect and challenge the environment in which they grow.

Four major concepts of mechanism of thermo tolerance

- Thermal dependence at biochemical and metabolic levels.
- Thermal tolerance in relation to membrane stability.
- Induced thermo tolerance through gradual temperature increase & HSP production.
- Photosynthesis and productivity during high temperature stress.

Thermal Stability of Cell Membranes

- Plasmalemma membranes playing vital role in cell functioning.
- Temperature causes a disruption of cellular activity or death.
- Heat injury often measured by ion leakage.
- Heat injury result in membrane protein de-naturation or membrane lipids melting leads to membrane rupture and loss of cellular components.
- Membrane lipid peroxidation observed in high temperature - a symptom of cellular injury.
- But, enhanced synthesis of anti-oxidants by plant tissue increases heat tolerance.
- Relationship between lipid composition and incubation temperature has been standardised for algae, fungi and higher plants.
- In *Arabidopsis*, high temperature exposure reduces lipid content by 50 % and unsaturated/saturated fatty acid ratio reduced by 33%.
- Membrane saturated fatty acids increases and confers heat tolerance.
- An *Arabidopsis* mutant accumulates larger amounts of 16:0 fatty acids, increases saturation of chloroplast lipids.

Heat Shock Proteins (HSPs)

- During rapid heat stress, synthesis and accumulation of specific proteins ascertained, called HSPs.
- HSPs also occurring with gradual increase in temperature.
- Three classes of HSPs are available and they are called as HSP90, HSP70 and Low Molecular Weight (LMW) HSPs (15-30 kDa).
- HSPs induced by heat stress at any stage of development.
- Under severe stress, HSP70 and HSP90, mRNA increase by 10-fold & LMW HSPs increase by 200-fold.
- Three other proteins are also considered as HSPs (eg., 110 kDa polypeptides, ubiquitin and GroEL proteins).
- Arid and semi-arid crops synthesis substantial level of HSPs under stress.
- Under laboratory condition in cotton, high temperature ranging from 38 to 41°C induce the synthesis of HSPs.
- In field cotton, eight new polypeptides accumulated in 50% of dry land leaves with soil water deficit showing midday Canopy Temperature of 40°C.
- Molecular weight of polypeptides were 100, 94, 89, 75, 60, 58 and 21 kDa.
- In soybean also, similar HSPs accumulation observed greater in non-irrigated treatments.
- Protective role of HSPs further proved that mutants unable to synthesize HSPs.
- Cells in which HSP70 synthesis is blocked or inactivated, become more susceptible to heat injury occurs.

Role of Heat Shock Proteins (HSPs)

Plants have the capacity to interact with the environment in many different ways and to survive under extreme abiotic and perhaps also biotic stress conditions. The response to heat stress (hs) is highly conserved in organisms but owing to the sessile life style, it is of utmost importance to plants. The hs-response is characterised by (i) a transient alteration of gene expression (ii) by the acquisition of a higher level of stress tolerance (acclimation). The induction of HSP-expression is not restricted to high temperature stress, HSPs are also linked to a number of other abiotic stresses including cold, freezing, drought, dehydration, heavy metal, and oxidative stresses. HSP are molecular chaperones, which either prevent complete denaturation (small HSP: sHSP) or are supporting proper folding (other HSP) of enzymes under or after protein denaturing conditions. Manipulation of the heat stock protein response has the potential to improve common stress tolerance that may lead to a more efficient exploitation of the inherent genetic potential of agriculturally important plants. Another approach is expression of gene involved sugar alcohol like aspartate decarboxylase and glycine betaine that will be used for resistant against heat tolerant.

- *Arabidopsis* plants engineered to overproduce glycine betaine, showed increased thermo tolerance, suggesting that nitrogenous osmoprotectants can have a role in cellular protection against heat stress.
- Engineering tobacco plants to express bacterial aspartate decarboxylase, the enzyme catalyzing the decarboxylation of aspartate to beta alanine is another approach.

Mechanism of HSPs for Heat Tolerance

The exact mechanism is still not certain, but it is hypothesised that HSP70 participates in ATP-dependent protein unfolding or assembly/disassembly reactions, prevents protein denaturation. Under such condition, HSPs provide significant basis for heat tolerance in crops to combat global warming.

- LMW (low molecular weight) HSPs play in structural role for maintaining cell membrane integrity.
- Other HSPs are associated with organelles such as chloroplasts, ribosomes & mitochondria.
- In tomato, HSPs aggregates into granular structure in cytoplasm, protect machinery of protein synthesis.

Photosynthesis *vs* High Temperature

- Optimum temperature for photosynthesis vary among crops due to crop adaptation to wide thermal environment.
- Increase of 0.1 to 0.2°C not likely to have substantial impact on leaf photosynthesis.

- Crop adaptation (tolerance) to temperature increase would be advantageous and can be developed.
- In tolerant species, global warming may not affect leaf photosynthesis or of closed canopy.
- Most of the C_3 crops are more sensitive to high temperature which may affect the photosynthesis.
- In future climates, temperature optimum for photosynthesis is likely to increase with increase of CO_2.
- CO_2-induced crop yield increase will be more in warm than cool environments.
- Elevated CO_2 will increase photosynthesis, provided the temperature does not increase at an alarming level.

Crop Growth and Development in High Temperature

Cereal development has five major developmental stages which are sensitive to temperature :

- Germination - seeding to seedling emergence.
- Canopy development - emergence to tillering or vegetative spikelet initiation.
- Spikelet production - first spikelet initiation to terminal spikelet formation.
- Spikelet development - terminal spikelet formation to anthesis.
- Grain development - anthesis to maturity.
    - All development stages are sensitive to temperature.
    - Development increases with rise in temperature.
    - Linear function of daily mean temperature and crop growth occurs.
    - Linear model within 10 to 30°C, but non-linear model describes development at extreme temperature stress.

Vegetative Phase

- All genotypes are sensitive to temperature.
- Temperature sensitivity varies greatly with genotypes.

- Duration of phase from sowing to first spikelet initiation, less sensitive to change in temperature than other phases.
- Genotypes differ in thermo-tolerance during this phase.

Flowering Phase

- From first spikelet initiation or terminal spikelet formation until anthesis have grater impact on grain yield.
- Spikelet number and floral number (potential grain number) occur in this phase – highly sensitive to temperature.
- Grain weight less sensitive at moderate temperature rise.

Grain Development Phase

- From 25 to 35°C, mean grain weight is declined by 16% for each 5°C temperature rise.
- Grain yield get decreased by 17% for each 5°C rise.
- For every 1°C temperature rise, decrease in grain yield by 8 to 10%, due to reduction of 5-6% in grains formation and 3-4% in grain weight.
- Temperature affects translocation, lead to alteration in source and sink activities.
- Temperature affects the grain weight by reducing assimilate availability and mobility.
- Grain weight reduction by heat stress because of reduced duration of grain growth.
- Minimal influence of temperature occurs on grain weight at maturity.
- Differential response of grain starch & N accumulation occurs.
- Warming increases drymatter accumulation in all cultivars.
- Accumulation of nitrogen is even higher than drymatter.
- Sucrose to starch conversion in endosperm decreases at higher temperature.
- Mature grain weight significantly reduced at highest temperature. Heat stress associated with reduced levels of soluble starch synthase activity.

- At high temperature reduction of grain yield occurs due to :
  a) Reduced number of grains
  b) Shorter grain growth duration; and
  c) Inhibition of sucrose assimilation in grains.

Extreme Temperature Effects on Crops

- Two types of of extreme temperature stress affects the crops - Heat and Cold.
- Global rise in temperature maintain either or both of more frequent high temperature stress and less frequent cold temperature stress.
- Higher temperature extends effective growing season under cold temperature stress prone areas.
  - Shift essentially important for crops (eg. Rice) grown in tropical centres of origin and in temperate latitude during warm seasons.
  - Global warming has greater impact in northern than southern hemisphere due to larger cultivated areas in northern hemisphere.
  - Increased temperature affects crop calendar in tropics.
  - Reduces length of growing season, where more than one crop is grown per year.
  - In semi-arid regions with wide diurnal temperature variation markedly increases temperature injury.
  - In India, wheat shows an adverse effect on seedling establishment, accelerates early vegetative development, reduces canopy cover, tillering, spike size and yield.
  - High temperature in second half of February (25/10°C), March (30/13°C) and April (30/20°C) reduce viable floret numbers, and grain filling duration.
  - Temperature stress reduces yield in December / January sown wheat crop.
  - In Rajasthan, similar situation is observed for sorghum and pearl millet exposed to high temperature.

- After sowing, high temperature scenario (air & soil temp. >40°C; mid-day soil surface temp. >50°C) is quite common in summer sown tropical crops.
- Acute temperature affects anthesis (opening and functioning of flowers) in crops.
- In rice, heat stress at anthesis stage prevents anther dehiscence and pollen shedding, reduces pollination and grain numbers.
- It is inferred that in tropics, many crops already subjected to heat stress, and further increase in temperature would cause crop failure in traditional areas also.

Long-Term Temperature Effects

- Continuous warming temperature reduces grain growth.
- Rice yield will be decreased by 9% for each increase of 1°C temperature.
- More genotypic variations in most of the grain crops require more thermal requirement for vegetative growth than reproductive development.
- With long-term increase, grain filling duration may decrease.
- Intensive research is required in identifying the species of various crops for developing tolerance to high temperature for using as donor parents is alternate efficient way in the management of temperature stress of both high and low.

❑❑❑

Chapter 8

Ozone, SO_2, NO_2 on Crop Growth and Productivity

Ozone is a form of **oxygen**. Oxygen exists in three different forms in the atmosphere. Ozone is also an effective greenhouse gas.

Oxygen atoms (O) are the simplest form of oxygen. The atoms exist as separate units; much of the oxygen in the highest part of the atmosphere exists in this form.

Oxygen molecules (O_2) are the most common form in which oxygen exists in the troposphere (lower atmosphere), and is the form of oxygen we breathe.

Ozone molecules (O_3) are three atoms of oxygen bound together. Ozone is poisonous to breathe.

Without ozone, there would be no life on Earth. Ozone protects us by forming a shield against harmful radiation from the Sun, known as **ultraviolet (UV)** radiation.

Mode of Action of Ozone

Ozone is a widespread secondary photochemical air pollutant in many parts of Europe, North and Central America and Asia. Ozone is important not only because of its phytotoxicity but also because of its tropospheric concentration which has increased considerably during past 60 years and is likely to increase at an annual rate of 0.5-2.5%. The harmful effects of O_3 on various components of vegetation such as forest trees, agricultural crops and ornamental plants are now well recognised. When the effect is severe, death of plant organ may occur. Effects on cells and organs affect organism growth, which ultimately leads to yield reductions (product synthesis) and quality changes. The reaction of plants to O_3 depends on the ambient levels, the amount diffusing in the leaf interior and the plants autonomous and environmentally modified resistance. The absorption of O_3 to individual plants is the consequence of the chemical potential gradient between the atmosphere and the site of deposition either on foliar surfaces or cells of the leaf interior. The harmful effects of O_3 occur due to reactions under gas and liquid phases after their uptake in plants and further reactions products.

Ozone that has passed through leaf internal air spaces dissolves in the aqueous layer. The breakdown of O_3 in pure water produces hydroxyl (OH), peroxyl (OH_2) and superoxide (O_2) radicals, although the reactions proceed very slowly at neutral pH. In contrast to O_3, H_2O_2 formed due to ozonolysis dissolves very well in the water phase and can be transported through the membranes and circulated within the plant. Further reactions of H_2O_2 lead to the formation of other reactive oxygen species such as superoxide and hydroxyl radicals, which can initiate lipid peroxidation, a chain reaction destroying the membranes. The initial site of injury caused by O_3 and / or O_3 generated ROS is the plasma membrane, resulting in changes in permeability, fluidity, potassium (K^+) exchange via ATPase reactions and calcium (Ca^{2+}) exclusion. A change in membrane function leading to a rise of intercellular Ca^{2+} that would lead to alteration of all sorts of intracellular metabolism.

O_3 reacts in the substomatal cavity with hydrocarbons such as ethylene and some terpenes. This oxidative process (ozonolysis) also produces H_2O_2 and aldehydes in a humid environment. The internal air spaces within the leaf are a potential site for O_3 reactions and formations of toxic compound within the cellular spaces

The plasma membranes are protected by antioxidants such as hydrophilous ascorbate (Vitamin C), lipophilous á tocopherol (Vitamin E) and enzymes such as superoxide dismutase (SOD) and peroxidases (POD). The first detoxifying barrier represents the antioxidant system found in the cell (apoplasm + symplasm). Ascorbic acid (AA) is an integral weapon in the defence against ROS generated by O_3. Apoplastic ascorbic acid is an important first line defence against O_3 as it is maintained in a reduced state. In plant cells, the most important reducing substrate for H_2O_2 detoxification is ascorbate peroxidase, which uses two molecules of ascorbate to reduce H_2O_2 to water through a series of reactions known as the ascorbate- glutathione cycle.

An alternative mode of H_2O_2 destruction is via peroxidises (POD), which are found throughout the cells.

An increase in POD activity has also been reported as a response to higher levels of H_2O_2. Photosynthesis (Ps), a core function in the physiology of plants is most susceptible to O_3. Reductions in Ps have been widely reported under ambient field conditions at higher concentrations of O_3. Meta analytical analyses on several varieties of wheat, soybean and rice also found varying degrees of response of photosynthesis to O_3. O_3 caused reduction in level of RNA transcript for the small subunit of Rubisco and also decreased the expression of photosynthetic genes including Rubisco. O_3 leads to reductions in m RNA levels of both small and large subunits of Rubisco of wheat cultivars M 510 and Sonalika grown in open top chambers (OTCs) receiving ambient+ 10 ppb and ambient+ 20 ppb O_3.

Reductions in photosynthesis affect carbon assimilation, translocation and accumulation in different plant parts. Ozone leads to inhibitory effects on phloem loading. The exposure of foliage to O_3 resulted in an accumulation of carbohydrate in the source leaves and reduced translocation to distant sink. The pattern of photosynthate allocation directly affects the plant growth and reproduction. O_3 exposure reduced the available carbohydrates to roots in soybean, cotton and a number of other crops. Even, the changes in the supply of or competition for assimilates, or in the synthesis and distribution of hormones required for the successful seed development and maturation, may affect the yield in grain crops. Studies have shown that O_3 exposure reduced yield in grain crops by decreasing ear and pod numbers. Grain yield in cereals is reduced by the effect of air pollutants on rate and the duration of grain filling due to impairment of production in carbohydrates and translocation of assimilates from the source organs to the grains.

Exposure to air pollutants may also induce negative impact on reproductive processes. These include modulation of pollen or ovule maturation; changes in the timing, rate or number of flowers produced, effects on seed and fruit development, yield, seeds germinability and seedling vigour. The impact of O_3 is known to vary between species and cultivars to be influenced by both climatic factors and cultural practices. There is abundant evidence that current ambient O_3 concentration in many industrialized areas of the world are sufficient to cause significant yield losses in both agricultural and native species.

Ozone symptoms characteristically occur on the upper surface of affected leaves and appear as a flecking, bronzing or bleaching of the leaf tissues. Although yield reductions are usually with visible foliar injury, crop loss can also occur without any sign of pollutant stress. Conversely, some crops can sustain visible foliar injury without any adverse effect on yield.

Susceptibility to ozone injury is influenced by many environmental and plant growth factors. High relative humidity, optimum soil-nitrogen levels and water availability increase susceptibility. Injury development on broad leaves also is influenced by the stage of maturity. The youngest leaves are resistant. With expansion, they become successively susceptible at middle and basal portions. The leaves become resistant again at complete maturation.

Ozone depletion

Ozone depletion occurs when the natural balance between the processes of stratospheric ozone production and destruction is disturbed, resulting in more ozone being destroyed than is being produced. The main cause of ozone depletion is thought to be a group of compounds containing chlorine and bromine. These 'ozone depleters' act by removing one oxygen atom from the ozone molecule, so converting it into an oxygen molecule. Each atom of chlorine or bromine can destroy many thousands of ozone molecules in this way. Compounds which contain chlorine do not occur naturally in the upper atmosphere. However over the past 100 years, the amount of man-made compounds containing chlorine and bromine has increased, and this has led to a decrease in natural ozone levels.

Ozone 'hole'

The depletion of the ozone layer is not homogeneous across the globe. There has been negligible depletion in the tropics and large scale periodic depletion over Antarctica. Ozone loss is accelerated over Antarctica because of the local meteorological conditions. At the very low stratospheric temperatures during the winter, polar stratospheric clouds form, providing a catalytic surface for the reaction between chlorine and ozone. The photochemical destruction of ozone begins in spring with the return of sunlight. A circumpolar vortex of wind isolates the ozone depletion over the Antarctic, leading to the formation of the ozone hole. In summer, the circumpolar vortex breaks up and higher levels of ozone from lower latitudes mix with the depleted region above Antarctic, repairing the ozone hole. This does not lead to a complete absence of ozone, but rather a large reduction in its concentration over a wide area. This is the ozone 'hole'. In recent years, the ozone hole has covered an areas of about 23 million square miles, with ozone concentrations falling by 60%.

Ozone effects on Plants

Acute Effects

- Unregulated and programmed cell death
- Occurrence of lesions (it may occur within few hours of exposure)
- Flecking and stippling effects.

Chronic Effects

- Lesions may occur in a long span of time (lesions can be observed after few days to few weeks)
- If there is no lesions, definitely there will be senescence effect
- Bronzing effects and Chlorosis.

Ozone effects on plant processes

1. Germination

- Delayed germination in the seeds and racemes of oilseed rape plants
- Precocious germination in *Brassica campestris* and *B. napus*.
- Rice proteome with leaf injury (200 ppb)
- Decrease in germination ranged from 55 to 100%.

2. Pollen germination, flower initiation and reproduction

- Reduces pollen germination and/or pollen tube extension
- Reduced flowering in soybean and cotton
- Delayed and reduced flowering in watermelon and muskmelon.

3. Biochemical processes

- Reduction in CO_2 fixation by ribulose bisphosphate carboxylase.
- Inhibition of CO_2 assimilation.
- Reduced photosynthesis.
- Reduced assimilation and increased respiratory loss of CO_2.
- Reduced assimilate production and export.
- Reduced leaf area with less light interception.
- In nitrogen metabolic cycle, changes in amino acid and protein content.
- Seed protein content decreased in oilseed rape.

4. Plant growth

- Inhibited photosynthesis, carbon (sugar) production, altered carbon allocation to roots and stems, uptake of minerals (N, P, K and S).
- Advanced senescence and accelerated catalysis with rapid loss of protein and chlorophyll.
- O_3 impairs phloem loading and assimilate partioning.
- Higher shoot/root biomass ratio, lower harvest index (reduced grain filling period), altered leaf chemistry.
- Decrease root exudates and pore water dissolved organic carbon (DOC).
- Reduced leaf N content.

5. Nutrient Uptake

An important root- shoot interaction that might determine the overall response of plants is the ability of the root system to adjust nutrient acquisition capacity to meet variations in shoot demand caused by environmental changes. Plant roots can alter their nutrient acquisition capacity by adjusting their physiological, longevity, morphological and or architectural characteristics to meet changes in shoot nutrient demand. Active root nutrient absorption is a highly adaptive plant characteristic that influences acquisition of N and other nutrients in response to environmental factors.

6. Changes in Root-to-Shoot Ratios (RSR)

A shift in biomass allocation to the shoots (a decrease in root-to-shoot ratios, has been observed in a wide range of plant species in response to O_3 exposure. Reduced biomass allocation to roots may decrease nutrient uptake, as well as reducing the yield of economically important plant parts. Nutrient acquisition is facilitated by increased specific root length and increased fine root numbers.

7. Changes in Root Respiration and Carbohydrate Supply to the Roots

Root respiration is generally reduced by O_3 stress. Theoretically, such reduction can limit the energy supply required for active nutrient uptake. Root respiration is related to nutrient uptake. The decrease in translocation of C to the roots is viewed as the cause of decreased root mass. By decreasing both root growth and root respiration, it is likely that O_3 also limits nutrient acquisition from the soil environment.

SO_2 (Sulphur dioxide)

Sulphur dioxide is the product of the burning of sulphur or of burning materials that contain sulphur.

Mechanism of action

The response of stomata to SO_2 entry is largely dependent on leaf age, concentration and combination of pollutants. Low concentration of SO_2 stimulated stomatal conductance in *Vicia faba* L. within 15 mins of exposure, which persisted for several days. This has been attributed to the destruction of epidermal cells adjacent to stomata and accumulation of sulphur within guard cells. Once SO_2 enters through stomata, the route to the surface of a nearby subsidiary or epidermal cell is very short and therefore, the cells of the epidermis are more susceptible. The detrimental effects of SO_2 occur due to reactions under liquid phases after their uptake in the plants

SO_2 readily dissolves in the apoplastic water to produce mainly sulphite, bisulphite (HSO_3^-) and H^+ ions, reducing the pH of the medium. Most leaves have the capacity to detoxify sulphite and bisulphite, if the concentrations are not excessively high, by oxidizing them to less toxic sulphate ion. Several studies have shown that the disturbances caused by SO_2 to biochemical functions and cell structure of plants appear before visual symptoms or growth reductions.

The low concentrations of SO_2 have been shown to stimulate the growth and physiological responses, especially in plants growing in sulphur deficient soil, where SO_2 might be metabolized to fulfill the demand of sulphur as nutrient. However, the higher uptake of SO_2 turns toxic and is reported to damage plants and reduce growth and productivity by interfering with different physiological and metabolic processes.

The effects of SO_2 on physiological and biochemical characteristics of plants have been well documented. Physiological processes such as photosynthesis, respiration, stomatal activity, transpiration and translocation are reported to be adversely affected by SO_2. Alterations in various physiological functions were ascribed to changes in permeability of plasma membrane, by interfering with enzymatic activities, and altering metabolic functions and nutrient uptake and water relations. Photosynthetic pigments and many enzymes are associated with the membranes of chloroplasts. Aqueous SO_2 can cause damage to plant metabolism by acting as an electron transport system.

The inhibition of photosynthetic CO_2 fixation by SO_3^- and this effect was due to competition between CO_2 and SO_3^- for active sites of ribulose 1, 5 - biphosphate carboxylase. SO_2 affects photosynthetic pigments leading to breakdown of chlorophyll. Studies with soybean showed significant decrease in chlorophyll content and photosynthetic rate at 60ppb SO_2. Sensitivity of crops to SO_2 is furnished in Table 2.

Table 2. Sensitivity of crops to SO_2

Susceptible crops	Resistant crops
Lucerne	Asparagus
Barley	Cabbage
Buckwheat	Celery
Clover	Corn
Oats	Onion
Pumpkin	Potato
Radish	
Spinach	
Squash	
Tobacco	

SO_2 Injury symptom

Dicotyledonous: Bifacial, marginal and/or interveinal necrosis, chlorosis, necrotic areas can change in colour from white to reddish (or), brown to black.

Monocotyledonous: Injury symptoms starts at the tip of leaves and spread downward, Necrotic and chlorotic streaks with occasional reddish, pigmentation

- Exposure of succulent, broad-leaved plants to SO_2 and its by-product sulphuric acid usually results in dry, papery blotches that are generally white, tan, or straw-coloured and marginal and interveinal.
- In some species, chronic injury causes brown to reddish brown or black blotches. Both the upper and the lower leaf surfaces are affected.
- The leaf veins normally remain green. Chlorosis (Yellowing) and a gradual bleaching of the surrounding tissues is fairly common.
- Growth suppression, reduction in yield and heavy defoliation may also occur. Middle aged leaves and young plants are most susceptible to SO_2.

Effect of SO_2 on basic physiological processes

- Changes in permeability of plasma membrane by interfering with enzymatic activity and altering metabolic activity and nutrient uptake and water relations.

- SO_2 also cause in decreased total chlorophyll and xanthophylls contents in leaves.
- High amount of SO_2 decrease the level of leaf conductance, xylem tension and fructose level in plants.

Effect of SO_2 on photosynthesis

- Inhibition of photosynthetic CO_2 fixation by SO_2 and this effect was due to competition between CO_2 and SO_2 for active sites of ribulose 1,5-biphosphate carboxylase. SO_2 affects photosynthetic pigments leading to breakdown of chlorophyll.

Effect of SO_2 on morphological characters of plants

- SO_2 may possibly slow down the rates of division and expansion of cells, leading to higher percentage reduction in shoot length than root length.
- Due to inhibition in translocation of photosynthates results in greater loss in root biomass.

Biochemical and molecular effects of SO_2

- SO_2 interfere with the structure and permeability of cellular membranes and with enzymatic activity which affects the biochemical process.
- Sulphite and Bisulphite react directly with disulphide bridges of structural and enzymatic proteins causing a decrease in their functions.
- Over expression of SO_2 causes alterations in mRNA levels and glutathione peroxidase.

NO_2 (Nitrogen dioxide)

Nitrogen dioxide (NO_2), a highly reactive gas is formed in the ambient air through the oxidation of nitric oxide (NO). Nitrogen oxides occur in atmosphere naturally and as a result of human activities. Nitrogen oxides are formed mainly by the burning of fossil fuels. This results from a chemical reaction between atmospheric N_2 and O_2 in the presence of heat to form NO, which then reacts again with O_2 to form NO2. The rate of reaction is determined by the temperature of combustion. Thus, unlike SO_2, NO_2 is not a component of fossil fuel, but results from a catalytic reaction of heat with atmospheric N_2 and O_2 in the combustion process.

Anthropogenic emissions of NOx lead to majority of all nitrogen inputs to the environment. The major sources of anthropogenic NOx emissions are automobiles and power plants. NOx emissions in Asia increased about 2.5 times from 1980 to 2000. Monitoring of air quality in Indian cities conducted by Central Pollution Control Board (CPCB) has shown a trend of dramatic increase in NOx concentrations since 1990 especially in the metropolitan cities. NOx emissions over the Indian region are growing at an annual rate of 5.5% $year^{-1}$. Annual average NOx concentrations varied from 4.3 to 42.9 ppb in various parts of the country. In the present scenario, NO_2 concentration is critical in Howrah (eastern region) where annual mean concentration is 42.9 ppb while it was 31.8 ppb in 1990. In other parts of the country, however, NO_2 concentrations were reported within the prescribed standards. Annual mean concentration of NO_2 ranged from 10.1 to 31.3 ppb in different zones of Varanasi city, situated in eastern Gangetic plain of U.P in 1990, whereas the same varied from 16 - 155 ppb during 1999-2001.

Mechanism of action

The two oxides of nitrogen i.e. NO and NO_2 are toxic air pollutants, but NO is rapidly oxidizes to NO_2 in free atmosphere. The uptake of nitrogen dioxide by plants occurs predominantly by foliar deposition. The mode of entry for the majority of gaseous NO_2 is through the stomatal openings.The entry of NO_2 into leaves is similar to that of SO_2, however, entry through cuticle is higher as cuticular resistance against NO_2 entry is lower than for SO_2 or O_3. NO_2 after entering into the leaf, dissolves in the extracellular water of the substomatal cavity to form both HNO_2 and HNO_3, which dissociates to form nitrate, nitrite and protons. NO_2 and NO belongs to the rare odd electron oxides of main group elements. The ease of dimerization of nitrogen dioxide yielding di nitrogen tetraoxide shows that NO_2 is a free radical. After dissolving into the aqueous phase of the apoplast two reactions may occur: reduction by antioxidants such as ascorbic acid to produce nitrous acid (HNO_2) and dehydro ascorbate or dissociation to produce nitrate (NO_3^-) and nitrite (NO_2^-). NO_2 is also capable of initiating peroxidation processes within lipid membrane. At higher concentration, NO_2 can cause acute injuries on plant leaves, while chronic exposure to lower concentration of NO_2 reduces the growth of plants. Ultrastructural changes associated with invisible injury have often been linked to reductions in transpiration and photosynthesis.

Gaseous oxides of nitrogen often restrict the growth of plants. The effect of air pollutants on wheat and mustard crops was assessed and it was observed that higher load of pollutants (SO_2 6.5 ppb and NO_2 9 ppb) showed maximum reductions in growth and yield, ascorbic acid content and photosynthetic pigments.

□□□

Chapter 9

UV Radiation on Crop Growth and Productivity

Introduction

The sun is essential for sustaining life on the earth; however, sunlight also contains a small amount of short wavelength ultraviolet (UV) irradiation, which is harmful to life on planet earth. Fortunately, most of this harmful UV irradiation is filtered out by the stratospheric ozone layer, which strongly absorbs UV light. Unfortunately, this protective shield is being continually damaged by human activities. During the past eighty years, the ozone layer has been damaged by the release of ozone depleting substances (ODSs) such as chlorofluorocarbons (CFCs), hydro chlorofluorocarbons (HCFCs), methyl bromide (MeBr) and other industrial compounds containing halogens.

Ultra-Violet (UV) radiation is one form of radiant energy emitted by the Sun. The Sun emits a range of radiation across the electromagnetic spectrum; ultraviolet is the range between violet light and X-rays. UV radiation is subdivided into 3 categories: UV-A, UV-B and UV-C (Fig. 1).

UV-A is the least damaging form of UV radiation and reaches the Earth in the greatest quantities. This is the UV light responsible for the tanning and ageing processes of the skin.

UV-B is potentially very harmful, as it can result in melanoma and eye cataracts. Most of the Sun's UV-B light is absorbed by the ozone layer. The depletion of the ozone layer will allow more of the UV-B light to reach the surface of the Earth, resulting in more cases of skin cancer and other diseases, damage to cell tissues, and other environmental problems.

UV-C is the most damaging form of UV light. Fortunately, it is almost entirely absorbed by the oxygen and ozone in the stratosphere before reaching the surface of the Earth.

These three UV radiations cause environmental stress because the cell constituents like proteins and nucleic acids absorb UV radiation in the range of 250-400 nm (UVA and UVB) and cause death of the tissues.

The agro-ecosystem is unique in that most of it has been created and controlled by humans. The opportunities for mitigating the impacts of increased UV-B may be easier in such a system because it is likely to have fewer environmental variables to consider than natural systems and can be modified on an annual basis. Breeding crop varieties with increased

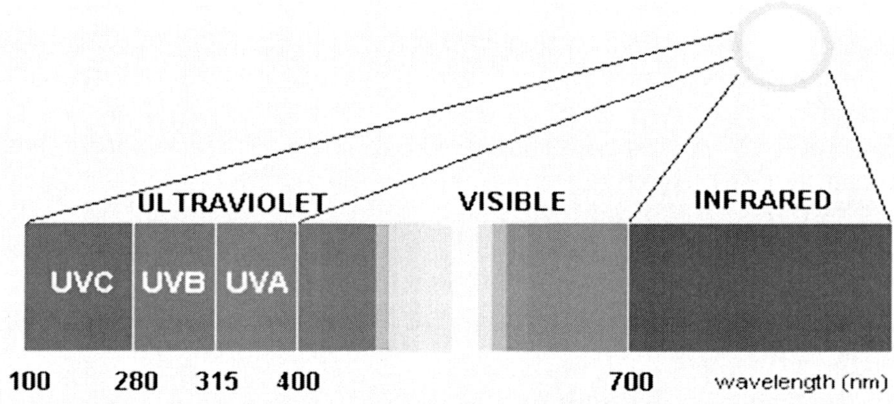

Fig. 1. The electromagnetic spectrum

resistance to UV-B, for instance, could be an effective mitigation strategy for agricultural systems.In plants, UV-B radiation damages the membrane surrounding the cell, and the DNA within the cell nucleus. Other primary targets are the membrane-bound chloroplasts, which use sunlight to convert carbon dioxide and water into sugar. With damaged chloroplasts, the plant's capacity to produce energy is reduced and plant yields eventually decline. Plants do have the ability to repair damage caused by UV-B radiation; however, when the disruption exceeds the plant's ability to repair itself, permanent damage occurs. UV-B radiation can affect a number of important plant processes and characteristics in the following ways.

- *Photosynthesis:* the ability of chloroplasts to produce sugars may be disrupted, resulting in reduced leaf productivity.
- *Growth:* plant height, leaf area and fruit size can be altered.
- *Reproduction:* exposed pollen and ovule can be damaged, leading to reduced fertility and seed production.
- *Ecology:* differential tolerances may favour the growth of one plant type over others, an important consideration in weed-crop situations and in pasture mixtures.
- *Quality:* surface blemishes on vegetables or fruits reduce the quality of the produce. The production of pigments in response to UV-B may affect the taste of the crop. A reduction in crop quality may be far more damaging than a reduction in yield.

I. Effect of UV-B on Morphological Features of the Plant

- Decreased plant height
- Affects the crop growth - First order effects. include leaf photosynthesis (photosystems, thylakoid and grana membrane integrity) and photomorphogenic systems (developmental rates) and decreased vegetative growth. Secondary and tertiary effects which altered crop growth and development, affects light interception that lowered canopy photosynthesis, reduced fruit numbers and retention, and finally, biomass and yield reductions
- Reduced branching
- Leaf area becomes less due to leaf cupping, leaf rolling
- Retarded growth of roots
- Change in leaf colour
- Low canopy of the plants
- Reduced flowering
- Grain size becomes less
- UV-B radiation may induce leaf differentiation and senescence processes via modification of leaf structure
- Plant sensitivity to UV exposure might be determined by direct damage to cell structural and functional elements or by ineffective acclimatization process

- Exposure to UV-B decreased plant height, leaf area and plant dry weight increased auxiliary branching and leaf curling
- A few weeks of UV-B exposure, leaf area and plant dry weight of rice were significantly reduced
- If the UV-B dosage exceeds the limits of tolerance, plant leaf anatomy is changed and biomass is decreased
- Enhanced UV-A decreased leaf area per unit plant biomass but increased biomass productivity both per unit leaf area and per unit leaf nitrogen
- Latest studies show significant UV-B radiation effect on barley growth parameters including stem height, sprout count, leaf area and biomass (Correia *et al.*, 1999)
- Increased solar UV-B lead to reduction in photosynthesis and production of some crop species
- Increased UV-B radiation induces a significant reduction in total biomass in a number of rice cultivars, accompanied by a reduction in tiller number and photosynthetic capacity of the plants.

II. Effect of UV-B on Physiological Processes of the Plant

1. Photosynthesis

- **Photosynthetic machinery:**
  - Inactivation of photosystem II (PSII)
  - Degradation of the D1 and D2 proteins of PSII
  - Decreased thylakoid membrane integrity
  - Reduced activity of Rubisco and other enzymes
  - Decreased levels of chlorophyll and carotenoids
  - Down-regulation of photosynthetic genes
  - Changes in chloroplast ultrastructure
- **Photosynthesis process:**
  - Due to UV-B, the double membrane of the chloroplast was disrupted and large vesicles are formed. This in turn reduces photosynthesis apparatus, hence the photosynthesis process decreased.

2. Respiration

Due to UV-B effect, less number of cristae in mitochondria was observed. Therefore reduces the respiration efficiency :

- Pigments are produced in the epidermis hence the quality of the produce is decreased
- Reproduction period becomes less
- Nodulation reduction and reduction in nitrogen fixation
- Transpiration rate increased
- Pollen germination decreased.

III. Effect of UV-B on Biochemical Features of the Plant

- Polyamines present in all living cells, displaying high biological activity and participate in diverse fundamental processes in cells. UV radiation triggers protective response in plants, including changes in activity of polyamines
- Flavonoids are synthesised
- Secondary metabolites accumulations are more
- Free radicals are produced
- Reduced ATP production
- Alteration in the plant hormones
- Increased ethylene concentration
- Lysosomes are formed
- Reduced sugars and starch
- Synthesis of Hydrogen peroxide
- Interruption in signal transduction
- Increased concentration of minerals, salts, etc.
- Destruction of proteins
- Destructive enzymes are synthesized.

IV. Effect pf IV-B at Molecular or Cellular Level

- DNA absorbs UV at the wavelength of 254 nm
- UV induced DNA damage causes pyrimidine dimer in the DNA, which interferes with DNA replication i.e., Thymine dimmers
- Leads to Mutations
- Affects the Transcription process
- Imperfect functioning of the cell division due to interfere in signal transduction
- Molecular chaperons synthesis is affected, hence protein folding will also be affected
- Plants are well adapted to minimizing damage that could be induced by AOS under natural growth conditions. However, oxygen toxicity emerges when the production of these AOS exceed the quenching capacity of protective systems due to environmentally adverse conditions such as drought, high light, water logging, high salinity, air pollutants, heavy metal toxicity and UV radiation
- Recent studies have shown that enhanced UV radiation alters membranes. This can be seen by the increase in malondialdehyde concentration (MDA), reduced monogalagtosyl diacylglycerol (MGDG), as well as an increase in ethylene and ethane concentration
- Protein degradation occurs under conditions of induced oxidative stress
- Some plants are more tolerant to UV than others because they produce a variety of secondary metabolites including flavonoids and anthocyanins. These compounds often accumulate in the upper epidermis cells of leaves and effectively absorb UV radiation thus preventing it from penetrating the leaf mesophyll cells
- Plants have developed a complex antioxidant system that includes reduced glutathione (GSH), ascorbic acid (ASA), - tocopherol, carotenoids and enzymes that protect plants against oxidative damage

- ASA is an ubiquitous soluble antioxidant in both plants and animals and is an important compound in plant cells, as well as being one of the most important reducing substrates for H_2O_2 detoxification
- The stresses imposed by UV-light irradiation can cause reactive oxygen species generation (ROS) such as $O_2$and H_2O_2
- Though H_2O_2 is an innocuous metabolite present in cells, irradiation with UV light breaks it down to extremely deleterious hydroxyl free radicals (OH)
- Since H_2O_2 can easily diffuse through cell membranes, it is extremely deleterious to cellular constituents such as DNA. Several studies have indicated that *in vitro* anthocyanins could act as effective antioxidants.

The effect of UV-B can be grouped into two broad groups i.e., Direct effects and Indirect effects.

Direct effects of UV-B on plants

- Changes in growth and morphology
- Reduced light interception
- Alteration in leaf anatomy and morphology
- Reductions in net carbon assimilation capacity
- Mechanistic damage to photosynthetic apparatus
- Changes in cell division
- Alterations in plant hormones or nucleic acids
- Change in biomass allocation and growth.

Indirect effects of UV-B on plants

- Susceptibility to pest and diseases attack
- Changes in competitive balance of plants and nutrient recycling
- Loss of membrane integrity
- H_2O_2 synthesis
- Depolorisation of cell's electric potential

- Affects signal transduction process leading to improper regulation of transcription
- Increased flavonoids synthesis
- Changes in cuticular wax.

Plants response or self protection of plants against UV-B

- Leaf curling
- Epidermal thickening
- Chloroplast thickening
- Changes in thickness of cuticular wax
- Synthesis of flavonoids
- Synthesis of anthocyanins
- Synthesis of alkaloids, waxes, cuticles
- Radical scavenging activity of polyamines.

UV Absorbing Compounds

- The responses to enhanced UV-B radiation vary among species and also cultivars in the same species. Accumulation of the UV-B- absorbing pigments is one of the ways by which plants alleviate the harmful effect of UV-B light
- Plants are considered to employ a variety of UV-B protective mechanisms, including enhanced accumulation of UV-absorbing compounds in the epidermis have been found to be involved in the protector of sensitive targets in leaves from the inhibitory effects of UV-B radiation
- A significant and positive correlation among the increase in the amount of UV absorbing pigment with UV-B exposure and the degree of UV-B sensitivity in cucumber cultivars
- Enhanced UV-B treatment typically induces an increase in UV-B absorbing compounds in leaf tissues
- Although cultivars of many crops have already been found to exhibit various degree of resistance to increased UV-B radiation, few studies have demonstrated a genetic relationship between

this protective mechanism and the resistance to UV-B radiation in cultivated plants

- A few investigations have reported that UV-B treatment increases the amount of UV-absorbing pigments in some rice cultivars
- Flavonoids are implicated as protective pigments in shoots and leaves exposed to UV-B light and their specific location in epidermal layer protects internal cell layers by attenuating the impinging UV-B radiation at the epidermis
- Other studies showed flavonoids are the largest UV-B absorbing compounds produced by barley. It has been shown that the photo-induced accumulation of these flavonoids is preceded by an induction of several enzymes of phenyl propanoid biosynthetic pathway.

V. UV Irradiation and Stress Tolerance

- Most studies on the effects of UV-B radiation on plants have focused on the potential adverse effects of high doses, photo inhibition, degradation of DNA or increased oxidative stress
- Exposure to lower doses of UV-B radiation, however, it may not impair growth or productivity in many plant species, there is evidence that exposure of plants to near-ambient doses of UV-B radiation can impart high-light and drought tolerance in woody species, including *Pinus pinea* L. and *Pinus halepensis*
- UV can in some cases reduce drought stress in plants and increase plant production through several potential water conservation and stress tolerance, although the interactive effects of the two stresses have also been reported to be neutral
- Drought tolerant species may sometimes also be more tolerant of UV radiation.

❑❑❑

Chapter 10

Climate Change on Photosynthesis and Respiration

Hundreds of experiments over the past 50 years have demonstrated that for nearly all C_3 crops photosynthesis increases as atmospheric CO_2 is raised. Elevated CO_2 (i.e. 700 ppm) stimulates leaf-level photosynthesis on the order of 20 to 60 percent, al-though this response varies considerably among crop species and with experimental methods.

1. Effect of elevated CO_2 on photosynthesis in C_3 plants

- Rubisco is CO_2-limited at current atmospheric CO_2 because it has such a low affinity for CO_2 (C_3 plants).
- Elevated CO_2 leads to higher CO_2:0_2 within the chloroplasts, which reduces photorespiration (C_3 plants).
- Elevated CO_2 appears to decrease the nonphotochemical dissipation of light energy within the chloroplasts of well-fertilized plants, thereby alleviating photoinhibition.
- Elevated CO_2 increases WUE and decreases water stress, which can contribute to higher rates of photosynthesis (C_3 and C_4 plants) over time. On the other hand, elevated CO_2 reduces stomatal conductance which places a limit on CO_2 transport from the atmosphere to chloroplasts, but the increase in CO_2 per se usually more than fully compensates.

Acclimation of photosynthesis to elevated CO_2 has been attributed to several mechanisms. Perhaps the most widely held view is that acclimation is linked to carbohydrate accumulation and its effects on expression of genes encoding photosynthetic enzymes. Presumably, carbohydrate production in CO_2-enriched atmospheres often exceeds.

If indeed photosynthetic acclimation is driven by accumulation of carbohydrates, then factors that limit sink activity, such as growing plants in small pots constraining roots or growing plants with insufficient nutrients to support new growth, should exacerbate the acclimation response. This is precisely the case, as less photo-synthetic acclimation occurs when plants are grown in large containers with plentiful nutrients compared to plants growing in small containers with limited nutrients. This may mean that much of the scientific literature on photosynthetic down-regulation in elevated CO_2 is of limited sig-nificance to crops because they are not grown in pots, and many (though not all) crops are fertilized to reduce nutrient limitations

Acclimation to high CO_2, however, is probably not completely explained by down-regulation of photosynthetic genes in response to inadequate sink strength. Photosynthetic acclimation to elevated CO_2 has even been observed in plants with very large sink ca-pacity, such as potato.

Effects of CO_2 enrichment on processing of carbohydrates in source leaves themselves can also lead to carbohydrate accumulation independent of sink activity. Starch accumulation, for instance, is sometimes a direct result of depressed photorespiration or may also follow from up-regulation of key enzymes in the pathway of starch synthesis, such as ADP-glucose pyrophosphorylase. Furthermore, a buildup of carbohydrates in source leaves could also follow from changes in export of triose-Ps fromchloroplasts or from effects on sucrose synthesis, phloem loading, phloem transport, or phloem unloading. These possibilities have not been well explored, especially in crops.

2. Effect of elevated CO_2 on photosynthesis in C_4 plants

Since the photosynthetic machinery of C_4 plants is saturated at relatively low concentrations of atmospheric CO_2, photosynthesis in C_4 crops such as maize, sugarcane, and sorghum is less affected by elevated atmospheric CO_2 compared to C_3 plants. It was even suggested that rising CO_2 might decrease the efficiency of C_4 photosynthesis if the production of C_4 acids exceeds the capacity of rubisco to catalyze the carboxylation reaction. Under such conditions, a greater proportion of CO_2, generated via decarboxylation of C_4 acids within bundle-sheath cells, may leak out into the mesophyll (24 percent and 33 percent of CO_2 leaked out of bundle-sheath cells of sorghum grown at 350 and 700 ppm CO_2, respectively.

Despite predictions that C_4 photosynthesis should be relatively in-sensitive to CO_2 enrichment, C_4 crops often realize significant benefit from elevated CO_2. There are several possible explanations for this. For example, C_4 leaves, before they have fully developed bundle-sheath anatomy, may largely function as C_3 leaves so elevated CO_2 could stimulate early photosynthesis, and therefore growth, of C_4 leaves beyond what would be expected from the effect of elevated CO_2 on C_4 photosynthesis. Early stimulation of leaf growth would accelerate canopy development and consequently in-crease integrated seasonal PAR absorption.

Greater than expected productivity of C_4 crops in CO_2 cnrichcd atmospheres might also be caused by increased WUE and alleviation of water stress. Stomates of C_4 plants are sensitive to atmospheric CO_2 in a fashion analogous to C_3 plants;as CO_2 increases, stomatal conductance decreases, which decreases the amount of tran-spiration per molecule of CO_2 assimilated.

Warming on Photosynthesis

The temperature at which photosynthesis is most rapid varies considerably from crop to crop, and therefore generalizations concerning the effects of global warming on photosynthesis itself are hard to make; but grouping crops according to the climates of their native habitats, seasonal "preferences," or photosynthetic pathways is a use-ful starting place.

Photosynthesis of crops originating in cool climates is greater at lower temperature ranges compared to crops that have their geographical origins in warmer climates. Crops with the C_4 photosynthetic pathway typically have higher temperature optima than C_3 crops. Photosynthesis of wheat (C_3), for instance, functions best at temperatures between 15 and 30°C while photosynthesis of maize (C_4) is most rapid when temperatures are between 30 and 40°C.

Temperature preference of C_3 and C_4 plants can be explained in part by differences in quantum yield. At low temperatures, quantum yield for C_3 plants exceeds that for C_4 plants because of the extra metabolic costs (2 ATP) associated with transporting CO_2 from mesophyll cells into bundle-sheath cells in C_4 plants.

Many crops are capable of substantial photosynthetic acclimation to high temperatures when they are imposed gradually. This could be an important aspect of crop responses to warming considered over the course of a cropping season. Acclimation may be linked to protection of temperature-sensitive photosynthetic components by heat-shock proteins, altered expression of genes encoding components of the photosynthetic machinery, or through homeoviscous acclimation of chloroplast membrane systems.

Interactive Effects of Warming and Rising CO_2

The effect of temperature on photosynthetic rate is greater at elevated CO_2. This is pre-sumably related to a decrease in the ratio of photorespiration to photo-synthesis at high CO_2 .Under ambient (non-saturating) CO_2 increased photorespiration negates a potential in-crease in photosynthesis that might be expected based on the Q_{10} of the carboxylation reaction, but at high CO_2 photorespiration is in-hibited and the temperature response of photosynthesis increases. Based on a biochemical model, a 300 ppm increase in CO_2 causes a 5°C increase in the temperature optima of C_3 photosynthesis. Therefore, at least theoretically, the stimulation of photosynthesis by elevated atmospheric CO_2 will be enhanced by warming.

Effect of elevated CO_2 on respiration

Direct effects on respiration

There are several reports that elevated CO_2has a direct, inhibitory effect on a plant's respiration rate. A direct effect of CO_2 on respiration is an instantaneous effect of CO_2 per se rather than a respiratory response to an effect of CO_2 on another process, such as photosynthesis or growth.If elevated CO_2 does directly inhibit respiration (and this is presently unclear), several effects on growth and yield might result. If the elevated CO_2 mainly affected an inefficient or unnecessary component of respiration, the reduced rate of respiration could result in a savings of assimilate that could then be used for additional useful components of respiration and for additional growth.

Indirect effects on respiration

It is well-known that elevated CO_2 enhances photosynthesis and plant growth (at least in C_3 plants). Increased photosynthesis and growth also stimulate translocation. Elevated CO_2 should, therefore, result in greater whole-plant respiration supporting growth and translocation as well as respiration supporting ion uptake and nitrogen assimilation (as-suming that bigger plants contain more minerals and proteins, which they usually do). The resulting increase in plant size should in turn stimulate whole-plant maintenance respiration. Finally, elevated CO_2 often results in a higher proportion of nonstructural carbohydrates (i.e., reserve materials).

In addition to increased growth, elevated CO_2 can also cause lower protein concentrations, perhaps in part through "dilution" by increased nonstructural carbohydrate levels. This response might be expected to reduce both growth respiration and maintenance respiration per unit biomass, though not necessarily respiration on a whole plant basis, and there is evidence supporting these responses in several experiments.

❑❑❑

Chapter 11

Climate Change on Crop Phenology

Phenology is the study of the timing of recurring biological events in the animal and plant world, the causes of their timing with regard to biotic and abiotic forces, and the interrelation among phases of the same or different species. Leaf unfolding, flowering of plants in spring, fruit ripening, colour changing and leaf fall in autumn as well as the appearance and departure of migrating birds and the timing of animal breeding are all examples of phenological events.

Importance of phenology for Climate Change

The IPCC (Intergovernmental Panel on Climate Change) concluded in its Third Assessment Report in 2001 that many physical and biological systems, such as hydrology, glaciers and ice, vegetation, insects, birds and mammals, are already reacting to changing temperatures. The importance of phenology lays in it's effectiveness as a tool to monitor impacts of climate change on plants and animals. Some imminent effects on vegetation include:

- Range shifts towards the polar regions and higher altitudes.
- Changes in population density and composition of vegetation.
- Longer growing seasons.
- Earlier plant flowering, earlier breeding times, earlier egg laying in the year.

The last two maybe best delineated by phenology. The intervals at which such events occur are very closely related to climate and weather conditions, especially temperature in spring and summer. Unlike the change in range shifts or changing composition of ecosystems,which maybe confounded by other drivers, such as land use change or habitat fragmentation and temperature. Thus, phenology is probably the simplest and most cost effective means of observing the effects of changes in temperature, and consequently, phenology has become an important tool in global change research. The use of phenology as a biological indicator of climate change presupposes

- Precise quantitative analysis of changes in phenological time series.
- A known relationship with temperature and or precipitation.
- An analogous change in corresponding temperature and or precipitation series over time.

Seasonal growth

Elevated CO_2 and temperature in certain species suggests that pasture management may change in the future. The grazing season may be prolonged, but whole-season productivity may become more variable than today. This is shown by studies of perennial ryegrass where, in spring, increased leaf extension occurred in elevated CO_2 whilst in summer, it was reduced. In high temperature, it was reduced in both seasons. In elevated CO_2 × temperature, leaf extension increased in spring, whilst in summer it decreased. Many organisms are near their tolerance limits and some may not be able to persist under hotter conditions. Higher temperatures in arid regions with cold winters may mean spring growth occurs earlier. Water reserves gained during the winter may, in some cases, be depleted earlier.

The effects on yield and phenology

Through global warming, an anticipated increase in temperature can potentially have various effects, e.g. spikelet sterility in rice, reversal of vernalisation in wheat, reduced formation of tubers in potatoes and loss of pollen viability in maize. Flowering is a very important event in crop development, as it is a phase which is particularly vulnerable to environmental stresses.

It is thought that extreme temperatures are more important than average temperatures in determining plant responses. Crop yields are affected by net primary productivity and also by the phenology of crop development. Increased temperature can speed phenological development, reducing the grain-filling period for crops and lowering yield. Crop yields were greater under elevated CO_2, but warmer temperatures reduced the duration of crop growth and, hence, the yield of determinate crops such as winter wheat and onion; but the yield of carrot, for example, an indeterminate crop, increased progressively with temperature.

In terms of temperature, a 12-day period of high temperature stress close to anthesis reduced spring wheat root biomass from 141 to 63 g m^{-2} by the end of the elevated mean temperature period, whereas mean temperatures over the treatment period had no effect on either above-ground biomass or grain yield at maturity. Interestingly, it was increasing maximum temperatures over the mid-anthesis period which was related to a decline in the number of grains per ear at maturity. Grain yield and harvest index also declined sharply with maximum temperature. This study suggested that high temperature extremes may reduce yields considerably.

Elevated CO_2 can aid the recovery of plants from high temperature-induced reductions in photosynthetic capacity. Soybeans for 52 days under normal air temperature and soil water conditions at atmospheric CO_2 concentrations of 360 and 700 ppm, but then subjected them to an 8-day period of high temperature and water stress. After normal air temperature and soil water conditions were restored, the CO_2-enriched plants attained photosynthetic rates that were 72% of their unstressed controls, while stressed plants grown at ambient CO_2 attained photosynthetic rates that were only 52% of their respective controls.

In the same study, total biomass was 41% greater under elevated CO_2 than under ambient CO_2 but reduced by high temperature, water deficit and high temperature × water deficit under both CO_2 concentrations. At maturity, seed dry weight and number per plant under elevated CO_2 were increased by an average of 32% and 22% respectively compared with ambient CO_2. The same parameters were reduced after high temperature × water deficit by 29% and 30% respectively in ambient CO_2 and elevated CO_2. Seed filling was earlier under high temperature and high temperature × water deficit. The rate of change in harvest index was unaltered by CO_2, while it decreased under the combined effects of high temperature × water deficit. Seed number explained 85% of the variation in yield, but yield was also related linearly to photosynthesis during seed filling, suggesting both are important determinants of yields under stress.

❑❑❑

Chapter 12

Climate Change on Water Relations

Availability of fresh water constrains crop production more than any other resources. Some plants are well adapted for dry environments, but unfortunately field crops are generally not among these. Most crops use tremendous quantities of water; a single maize plant transpire more than 15 liters of water per week and a maize crop may transpire more than 1,325,000 liters per acre (0.4 ha) during the growing season. Worldwide, nearly two-third of all those water cycled from land to atmosphere in a year passes through living plants.

There are three explanations for the linear relationship between yield and crop water consumption:

1. Carbon and water vapour share a common diffusive pathway between atmosphere and the interior of plant leaves; consequently, stomates must be open, and water transpired, in order for crops to assimilate carbon from air surrounding leaves;
2. Both water loss and photosynthesis are driven by absorption of light; and
3. Evaporative water loss significantly cools leaves and canopies, which alleviates high temperatures tress.

The overall influence of environmental changes on water use by a crop will be manifested mainly as a change in total evapotranspiration. Evapotranspiration is the combined measure of plant transpiration, evaporation from soil, and interception losses. Interception losses represent rainfall that evaporates directly from external surface of vegetation or from soil–surface residues without ever entering either soil or plant and can be significant in dense canopies. Environmental changes are also likely to affect leaf, plant, and crop water use efficiency (WUE).

I. Direct Effect of Climate Change

1. Effect of climate change on global hydrology

- It would be useful to predict the how climate change will influence the global hydrological cycle and regional precipitation patterns in order to understand and plan for potential effects on agriculture.
- A change in global hydrology would influence availability of water resources and therefore food production.
- It will also Influence the ratio of precipitation to evaporation, thereby influencing water levels of lakes and runoff and flow of rivers.
- Based on the historical data, it was estimated that precipitation has already increased 5-10 percent during the past century due to more intense rainfall events and that this has contributed to more stream flow.
- One region will get more intense form precipitation and other will experience prolonged dry spell due to erratic climate pattern.

2. Effect of elevated atmospheric CO_2

- Plants grown with CO_2 enrichment (ambient + 360 ppm) in controlled environments with well- mixed air, stomatal conductance and transpiration typically decrease 20 to 90 percent.
- Reduced transpiration coupled with enhanced photosynthesis, enable most of species to grow bigger and produce more yield per unit of water transpired by leaves. So that water use efficiency will eventually increase.
- Transpiration decreases (because of CO_2 - induced stomatal closure), would increase the canopy temperature due to dry air around leaf increase, the driving force for evapotranspiration from the canopy.
- A doubling of atmospheric CO_2 level will have the negligible effects on total water consumption by crops (total ET could change by -0 to + 10 percent). Crop productivity per unit of water consumed, however, is likely to increase 20 to 30 percent, suggesting that dryland agriculture may benefit and demand for irrigation water may drop.

- There are likely to be interactive effects on crop water use for plants subjected to multiple environmental changes, however for instance, warming is likely to cancel out some proportion of the increase in WUE associated with rising CO_2.
- Elevated CO_2 concentration leads increase the photosynthetic activity of C_3 plant and thereby increases the WUE.

3. Effect of elevated temperature

- The water vapor content of saturated air increase exponentially with rising temperature. The water content of air in intercellular space of leaves is generally very close to saturation regardless of temperatures. So the water density gradient between crop canopies and atmosphere will rise because of increase in the water density of air within leaves. Temperature increases causes stomata to open, increase water loss.
- Global warming will drive canopy and landscape scale evapotranspiration upward. In an experiment on soybean over the temperature ranges of 28-38°C, for instance, transpiration increased by ~4 percent for each degree C.
- Global Warmingis also expected to decrease WUE for both C_3 and C_4 crops. In the case of rice grown ambient and elevated CO_2 (700 ppm) in a temperature gradient tunnel, CO_2 -enrichment enhanced WUE by 40 to 50 percent between 24 and 26°C, but that increase in WUE dropped to ~20 percent at 30°C.
- Soil warming could have significant effect on root growth and function, thereby impacting water relation. In general, at the high end of optimal temperature range, high soil temperatures increase root hydraulic conductivity and thus water flux compared to cooler soil.

4. Effect of elevated atmospheric ozone concentration

- Atmospheric ozone enters plant through open stomates. Therefore, the effect of ozone on plants is linked with stomatal conductance values.
- In general, environmental factors that induce stomatal opening increase ozone susceptibility and factors that induce closing decrease susceptibility.

- The influence of ozone on stomates conductance (water relation) is complex and depends on the duration severity of expoture.
- Ozone reduces the hydraulic conductivity of roots by affecting root growth and whole plant eg., cotton which could limit the water uptakes as well as nutrient acquisition.

5. Effect of salinity

- Crops grown in a saline soil are sometimes unable to acquire enough water to meet their needs, even when soil-water contents are high. This is because of soil, plant, and atmosphere continuum (SPAC).
- Excess of solute reduces the water potential that makes less favorable to roots.
- Excess solutes cause toxicity to the membrane that reduces the membrane integrity and damage to other cellular contents and cellular metabolisms.
- Under stress condition, plants accumulate some osmolytes like sugars, amino acids and salts to protect against osmotic effect. Accumulation of osmolytes inhibits the growth and yield of the crop.
- High temperature hasten the evapotranspiration rate will deplete the soil moisture and photosynthetic activity of crop plants

II. Indirect Effects of Climate Change on Water Relations

- Change in the balance between precipitation and snow fall.
- Increase in atmospheric temperature that trigger evapotranspiration from surface soil and leaf surface leads to depletion of soil moisture.
- Unequal precipitation pattern.
- Melting of glacial ice which leads to increases in sea level may resul in

  a) Flooding of low lying area

  b) Erosion

  c) Salinity and alkalinity problem

d) Reduction in water quality

e) Impact on crop production

- Change in cropping pattern and
- Elimination of seasonal bound crop.

❒❒❒

Chapter 13

Climate Change on Crop Yield

Introduction

Climate change variables including precipitation (amount and distribution), temperature and atmospheric CO_2 concentrations are expected to alter agricultural productivity patterns worldwide. Carbon dioxide is a plant nutrient, and atmospheric enrichment has the potential to enhance plant productivity. At least in some regions, agriculture may be one of the brightspots, 'the silver lining in the climate change cloud'. But higher global temperatures and altered precipitation patterns are expected to accompany the higher CO_2 levels, and these factors may lessen any production increases or even depress production below current levels. The myriad of modeling studies attempting to project the short- and long-term impacts of climate change on agriculture are consistent only in highlighting that the nature of the productivity change itself will vary. Realized yield changes will reflect differences in local environments as well as differences in access to seed and management technologies that may offset negative climate change impacts.

Regardless, with any potential changes in agricultural productivity comes a potential for associated changes in crop nutrient use. Local potential yield levels are determined by prevailing climate, ambient CO_2 and crop characteristics, but these yields are almost always limited by root zone resources such as nutrients and water and further reduced by pests and diseases. The interactive effects of soil moisture and nutrient availability are two key edaphic factors that determine crop yield. The current nutrient management recommendations are based on an understanding of crop-specific needs for achieving expected yields and soil-specific nutrient supply characteristics. To what extent does our existing knowledge remain useful undera changed climate? Addressing this question requires anassessment of the potential for global climate change factors to influence the physiological efficiency (PE) of nutrient use within the plant and to alter the availability of nutrients in soil and their transport through soil and across root membranes.

Principal climate change drivers in agriculture

Five main climate change-related drivers will affect the agriculture sector in ways that will vary in intensity and importance across the regions. They are: temperature rise; precipitation patterns, including rainfall and snow; the incidence of extreme events (floods and droughts); sea-level rise; and the atmospheric carbon dioxide content.

At high latitudes, crop yields are expected to rise with temperature increases of 1-3 ° C, but fall, due to declining crop health, once 3° C is exceeded. At lower latitudes, crop yields are expected to decline with temperature raises as little as 1–2° C. Overall, the benefits of carbon dioxide enrichment on photosynthesis are likely to be outweighed by increased temperature and lower rainfall. It is expected that agriculture (without any further adaptation), especially in the dry and wet tropics, will be more affected by an increased frequency of extreme events, rather than the mean change in climate. Additionally, there is anticipation of greater fire risk and incidence of pests and diseases.

Climate change impacts on crop yield

The changes in crop production related climatic variables will possibly have major influences on regional as well as global food production. The likely impacts of climate change on crop yield can be determined either by experimental data or by crop growth simulation models. To predict future impacts on crop yields, crop models present valuable approaches. A number of crop simulation models, such as CERES-Maize (Crop Environment Resource Synthesis), CERES-Wheat, SWAP (soil-water-atmosphere-plant),InFoCrop have been widely used to evaluate the possible impacts of climate variability on crop production, especially to analyze crop yield-climate sensitivity under different climate scenarios.

If water availability is reduced in the future, soil of high water holding capacity will be better to reduce the frequency of drought and improve the crop yield. With climate change, the growing period will reduce, and the planting date also needs to change for higher crop production. Climate change can decrease the crop rotation period, so farmers need to consider crop varieties, sowing dates, crop densities and fertilization levels when planting crops. Quantification of climate uncertainty is an important indicator for crop yield variation in future climate scenarios.

Crop responses to climate change

Crop growth, development and yield responses to climatic variability area mixture of linear and non-linear functions. Changes in the mean, variability and rate of occurrence of extremes in temperature all affect crop processes but not necessarily the same processes. Photosynthesis and respiration can change continuously and non-linearly in response to incremental increases in temperature, but short periods of high temperatures can do disproportionate damage when coinciding with flowering or pollination. Likewise, mild water stress has a different effect than prolonged drought or flooding.

Effects of elevated CO_2 on nutrient partitioning and grain yield in rice

Relationship between plant growth, grain yield, nutrient acquisition and partitioning in rice (*Oryza sativa* L.) under elevated $CO_2$was observed. Plants were grown hydroponically in growth chambers with a 12-h photoperiod at either 370 or 700 μmol CO_2 mol"[1] concentration. Plant dry mass (DM), grain yield and macro and micronutrient concentrations of vegetative organs and grains were determined. Elevated CO_2 increased biomass at tillering, and this was largely due to an increase in root mass by 160%; whereas elevated CO_2 had no effect on total nutrient uptake (N, P, K, Mg and Ca). However, nutrient partitioning among organs was significantly altered. N partitioning to leaf blades was significantly decreased, whereas the N partitioning into the leaf sheaths and roots was increased. Nutrient use efficiency of N, P, K, and Mg in all organs was significantly found increased at elevated CO_2. At harvest maturity, grain yield was increased by 27% at elevated CO_2 while grain (protein) concentration was decreased by a similar magnitude (28%), suggesting that critical nutrient requirements for rice might need to be reassessed with global climate change.

Effects of elevated CO_2 on spring wheat

To study the effect of increased atmospheric CO_2 on crop growth and dry matter allocation in spring wheat, pot experiments have been carried out in two glass houses at ambient and doubled CO_2 concentration. Wheat plants were grown at different supplies of N, P or K.

The results revealed that doubling of ambient CO_2 resulted in a large increase in total biomass (+70%) and grain yield when the nutrient supply was optimum. With strong N and K limitation, this CO_2 effect was about halved and with strong P limitation, it became almost nil. Doubling of ambient CO_2 resulted in a 10% lower minimum N concentration in plant tissue and in no change in the minimum P concentration.

❑❑❑

Chapter 14

Climate Change on Seed Yield and Quality

Human beings depend on seeds (grains) and fruits produced by plants, with about 75 percent of world food supply coming directly or indirectly from the seeds of just three crops: wheat, maize, and rice (Egli, 1998). Biological grain yield is the total mass of all grain produced by a crop per unit of ground area. Harvested yield may be considerably less than biological yield due to various losses before and during harvesting. Both biological and harvested yield are affected directly or indirectly, by environmental changes. Grain quality includes physical and chemical properties (e.g., types and amount of protein, carbohydrates, and lipids) of harvested seeds. Seed composition varies considerably among varies crop species. Cereals seeds are generally high in carbohydrate and low in proteins and lipids. In contrast, pulse crops are high in proteins and have relatively low levels of carbohydrates and lipids, whereas oilseeds have particularly high concentration of oil in their seeds. The environmental changes that affect the crop physiological process have potential to ripple through subsequent growth and development ultimately affecting both grain quantity (biological and harvest yield) and quality. Early seedling growth and development affected by environmental changes, following phenostage may be affected, and that in turn can alter the next developmental change, and so on, even spanning multiple crop generations. A change in yield brought about by environmental change could be accompanied by a different change in quality. For example, wheat grain and grain protein concentrations are often negatively correlated. So an environmental change that enhances the yield may reduce grain protein vice versa.

Sexually Reproductive Development

The switch from vegetative to sexually reproductive development is a defining movement of crop grain crops. During that transition, shoot (apical) meristems begin to built flowers (i.e., meristems form flower primodia), which eventually produce grain. Flowers are determinate organs in which male and female gametes (sperm and egg, respectively) are produced, mechanism facilitating union of male and female gametes (pollination and fertilization) are complex. They are also disrupted by unfavorable climatic conditions. In most of the modern crop varieties, the accumulation of a genetically related quantity of "heat sums" or "degree days" provide stimulus to initiate sexually reproductive development. In fewer crops, the onset of reproductive development is triggered by photoperiod (actually, night length), a phenomenon called photoperiodism. In other crops, a combination of accumulated heat units and night length serve as the cue to initiate the flowers.

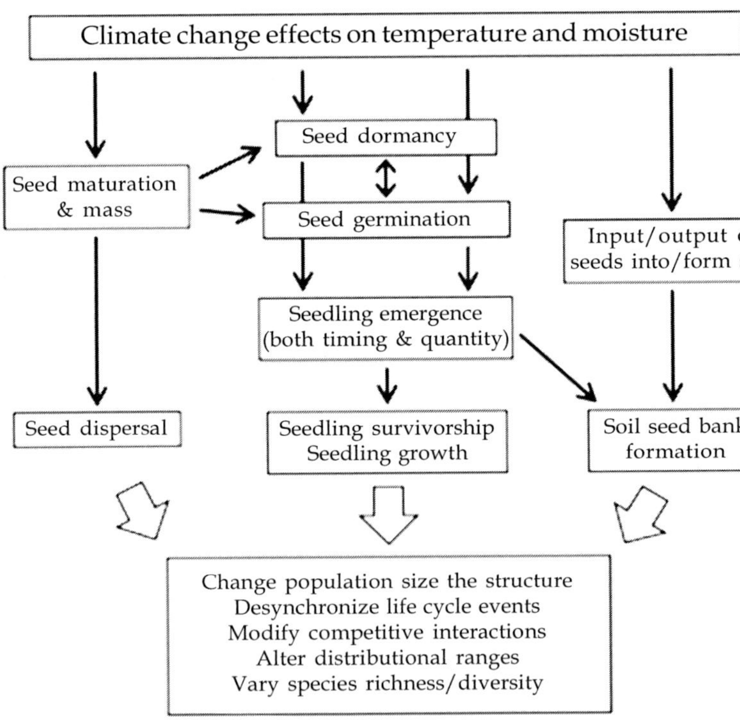

Fig. 2. Flow chart showing the effects of climate change on plant regeneration from seeds and repercussions on the early life history events of plants at the population and community levels.

Factors affecting seed growth

Seed growth is the complex process influenced by many factors (Fig 2). Photosynthesis, nutrient and water acquisition, translocation of raw materials into developing seeds, and conversion of precursors into storage compounds by the seed itself all contribute to seed size and quality. Seeds of most crops are able to synthesize a diverse array of complex organic molecules from a few simple precursors.

Ovary position within an ovary is important to both seed development and growth. Although ovules at the stylar end of the locule are first in line for fertilization, ovules at opposite end may be first in line to receive substrates of growth from mother plant. This effect can be striking in maize, where the kernels at the distal end of the cob often do not receive sufficient substrates to develop fully. So, depending on whether an environmental change affects pollen success or supply of substrates, the fertilization and growth of different ovule may be affected differently.

Yield component and variation

- Variation in yield of grain crops is caused by differences in seed number per unit ground area and/or differences in mean seed mass.
- Seed number and mean mass of individual seeds are often referred to as yield components.
- Final biological grain yield is the product of final seed number per unit ground area, the mean seed filling duration (SFD), and mean growth rate during seed filling (i.e., mean seed filling rate (SFR)).
- By definition mean seed mass equals SFD x SFR. To understand the effects of environmental change on yield, effects on each component of yield must be known.

Seed number

Final seed number per unit of ground area is the number of seed set (i.e. number of ovules fertilized) minus the number that are later aborted. The number of ovules capable of becoming mature will depend on canopy photosynthesis before and during flower growth. Conditions associated with high rates of photosynthesis, such as high soil nitrogen levels, abundant light, and high atmospheric CO_2, almost always result in a greater number of ovule. Conditions that limit the canopy photosynthesis, whether through effects on leaf area or biochemical aspects of CO_2 assimilation, are generally associated with fewer ovules. Seed set is especially sensitive to environmental conditions around the time flowers develop because that is when pollen is produced, young ovules are formed, and pollination and fertilization occur.

Seed filling duration

Seed filling is the time between the initiation of seed filling and initiation of seed dormancy. Seed filling can overlap with vegetative growth in indeterminate crops, but in determinate crops, initiation of seed filling generally corresponds to cessation of vegetative growth. Changes in SFD can cause substantial changes in final seed size and yield.

- Critical regulator of SFD is temperature, in many crops is inversely related to yield and temperature decrease increases SFD (Evans, 1975).
- Control of SFD may also be exerted through leaf longevity, water relations of growing seed (Egli, 1998).
- Supply of growth substrates in particular, as it is influenced by leaf-area duration, may play a role in controlling SFD.
- Leaf senescence may play a role in triggering cessation of seed growth and initiation of seed dormancy.
- Water potential of matured seed declined, and retain constantly in different crop species triggered seed dormancy (Ahmadi and Baker, 2001).
- Physical restriction seed growth may be exerted by the seed coat, carpels, cotyledon or endosperm cells, or other structures.
- Water stress, nutrient limitations, and high temperature, all lead to smaller seeds, and limited yield in most grain crops.

Seed filling rate

- Seed–filling rate is the average rate at which seeds accumulate bio-mass during the seed-filling period. Seed-filling rate can be strongly affected by temperature.

- SFR is usually affected by the supply of raw materials from the mother plant; it is sensitive to the number of endosperm or cotyledon cells available for expansion. Because cell number is set by the environmental conditions that prevail during the period of embryo and endosperm cell division.
- High temperature affects the potential for filling of individual seeds. Several reports for wheat indicated that large seeds are associated with large number of endosperm cells (Wardlaw, 1990).

Environmental changes

Elevated atmospheric CO_2 concentration

- It has been well known that increased CO_2 concentration in the atmosphere will increase the WUE efficiency of crop by reducing the transpiration rate in terms of by reducing stomatal conductance.
- Elevated CO_2 is also beneficial to reproductive growth that CO_2 has been pumped into glasshouses for many years to improve flowering characteristics (including quality and number) and fruit yields of several horticultural crops.
- Effects of elevated CO_2 on grain quality vary greatly among varieties of same crop species. Yield enhancements of nine soybean cultivars resulting from CO_2 enrichment ranged from 20 to 90 percent. For all crops, the positive effects on yield of a given ppm increase in CO_2 are greater at sub ambient $[CO_2]$s compared to super ambient $[CO_2]$s.
- Most of the experiments involved $[CO_2]$s from present ambient values up to about 700ppm, and yield of many crops is stimulated by about 1000 to 1500 ppm. Effect of elevated $[CO_2]$s on C_3 plants is similar to wheat. $[CO_2]$s level greater than 1000 to 2000 ppm can be deleterious to yield.

Components of increased yield

- Many factors can contribute to enhanced yield with CO_2 enrihment. Through its stimulatory effects on both photosynthesis per unit leaf area growth (or LAI), elevated CO_2 results in increased canopy photosynthesis, at least in C_3 crops. This enables crop to grow larger, producing more flowers and fruits. Ovule and seed number are positively correlated with amount of canopy temperature.

- In rice, a 47 percent increase in yield, panicle per plant, rice tiller initiation and number of potential seeds per plant, was stimulated by elevated CO_2 during two weeks following germination.

Grain quality

- A general response of plants to elevated CO_2 is a decrease in nitrogen concentration of vegetative organs. Since nitrogen transport to developing grains depend largely on amino acids exported from leaves, grain nitrogen (protein) concentration might decrease as well, affecting the quality of grains used in some food products.
- Change in grain nitrogen content is species specific. Grain nitrogen concentration was unaffected by elevated CO_2 in rice and soybean, reduced slightly in cotton, and reduced significantly in barley and wheat.
- In Wheat and also in other crops, reduction in grain protein concentration caused by elevated CO_2 can be minimized by application of extra nitrogen fertilizer.
- Elevated CO_2 increased the grain carbohydrate content in many cereal crops, make up about 80 percent seed mass. Increases in carbohydrate in relation to protein increases the malting quality of barley and cooking quality in rice.Elevated [CO_2] had only small effects on wheat-grain lipid concentration and composition.

High temperature on seed set and Number

Yield can be especially sensitive to high temperature around the times of anthesis, often seen through effects on final seed number. Flower and ovule number, pollination, pollen tube growth, and fertilization can all be reduced by high temperature, with consequent negative effects on yield.

- Pollen may be the most sensitive of all reproductive structures. Pollen viability is rapidly reduced when temperature exceeds about 30 and 34° C for wheat and rice, respectively.

- High night temperature is detrimental particularly in cowpea. Night temperature above 30°C leads to low pollen viability and failure of anther to dehisce. Loss of pollen viability is generally most severe when high temperature occurs five to ten days before anthesis.
- High temperature during pollen development can lead to shriveled, fewer, and smaller pollen grains. May involve premature degeneration of tapetal cells, tapetum mediates the transfer of material from mother plant to microspore. High temperature may also affect several enzymes that are important in carbohydrate metabolism. It appears that most damage occurs during microspore development. Sensitivity of pollen to high temperature may be lack of Heat Shock Proteins (HSPs) in pollen.
- High temperature can also affect the ovule function and cause excessive proliferation of nucellus or integument tissues, resulting in crowding of embryo sac, (these change may interfere with hormonal signaling from the embryo sac, which controls the direction of pollen tube growth).
- High temperature during very early stage of seed development leads to abortion of ovule. In wheat, grain set is reduced by exposing to above 30°C and can reduce the grain yield by 70 percent.
- Stigma and style are less temperature sensitive than anthers. For example in rice, upper temperature threshold for proper functioning of stigma and style is at least 6°C higher than pollen.

Seed filling duration, seed- filling rate and yield

High temperature during grain filling period reduce individual seed size at maturity, which may results from combined effects on seed -filling duration (SFD) and seed-filling rate (SFR). Effects of temperature on SFD may be most important to yield.

- In wheat, SFD decreased 3.1 days for each 1°C temperature increase. Decrease in SFR at high temperature are often attributed to specific effects on the enzymes involved in converting assimilates provided by the mother plant into seed mass.
- The activity of soluble starch synthase (SSS), a key enzyme in amylopectin production, is particularly sensitive to temperature

above 30°C. Amylopectin accounts for 50 to 60 percent of the seed mass in maize, rice, and wheat.

- High temperature during early kernel development (i.e., the first 10 to 15 days after pollination) can inhibit cell division and formation of endosperm tissue.
- Mostly high temperature generally reduces the yield of grain crops, For example, grain mass of wheat can decline 5 percent for each 1°C rice in air temperature from 17.7 to 32.7 °C and in rice grain mass can be reduced 4.4 percent for every 1°C temperature increase from 26.7 to 35.7 °C.
- For maize, it was estimated that a 6 °C warming during grain filling resulted in a 10 percent yield loss, and for sorghum, increased night time temperature (above an optimum) can have deleterious effects on yield. For every 1 °C temperature increase there will be 11 percent loss in yield in beans.

Seed quality

Seed quality is, however, sensitive to temperature during the seed-filling period. Because high temperature can differentially affect the various process involved in seed filling (for example, protein compared to carbohydrate synthesis, SFD compared to SFR), warming can affect seed composition. In addition, because seeds are often smaller at high temperature, their milling quality can suffer.

- Grain protein concentration increases in many crops (eg. soybean, bean, maize, rice, wheat, and barley) with warming. In wheat high protein correlates with dough strength; it is important bread making trait. Dough strength improves with warming up to a threshold and then detoriarates. This threshold ranges from about 25 °C in sensitive wheat cultivar to about 35°C more heat tolerance varieties. In barley, even small increase in protein concentration decreases malting quality.
- Activation of HSPs in seeds exposed to high temperature, although providing protection to the developing seed against the heat damage, also reduces the bread making quality of wheat. Often, increase in grain protein concentration with warming are due to decreased starch content rather than increased protein content.

- Wheat grown at higher temperature often produces seed exhibiting a chalky appearance attributed to small starch granules surrounded by numerous air spaces. Similar chalky appearance was observed for rice grown at high temperaeture, as result of changes in both size and shape of starch grains. In rice, it was estimated that starch amylase content decreased 0.4 percent for every 1°Crice temperature during the period of anthesis.
- Effect of high temperature on oil content and oil quality have been documented for several oil crops, including flax, sunflower and cocoa. In sunflower, the ratio of linoleic:oleic acid is about 6:1 at 12 °C and approaches 1:1 at 28°C. Since sunflower is valued more for linoleic and oleic acids, high temperature can significantly decrease the quality of sunflower oil.
- High temperature reduces the lipid accumulation in wheat kernels to the extent that flour properties can be affected.

Direct effect of climate change on seed development and quality

- High temperature cause poleward shift in cultivation of given species or genotypes
- Earlier planting of a given species or genotype.
- Replacement of a current species or genotype with one better adapted to higher temperature.
- Increase in temperature cause rapid increase in metabolic activity there by reduces the grain filling duration that may lead to reduction yield.
- Longevity, viability, gerrmination and vigor of the seeds and seedling are affected by elevated temperature, CO_2, O_3 and salinity regime.
- Size and structure of the population changed.
- Desynchronization of life cycle of some species.
- High temperature cause pollen and spikelet sterility often reduce the seed set percentage

Indirect effects

- Affects the soil seed bank.

- Reduced seed maturation and seed mass often results in reduced weight of seed and yield.
- Reduced the seed vernalization process.
- Quantitative changes in the emerging seedlings and changes in the time of germination.
- Reduction in seed quality often associated with consumption preference and marketing.

Chapter 15

Climate Change on Crop Pest Interactions

Insects form one of the major components of agricultural biodiversity and are through to mainly have negative impact on agriculture as pests of crop plants, though their positive role as biocontrol agents and pollinations is of equal economic importance. Any change in the population dynamics and distribution pattern of insects will have a significant impact on crop productivity.

The following scenarios can be visualized regarding impact of climate change on pest dynamics in agriculture.

1. Climate change, on Indian Agriculture with an increase in concentration of carbon di oxide, the nutritional status of crop will change, and the net effect on agricultural production will depend upon interaction between pests and crops.
2. Gradual climate warming will lead to changes in the composition of pest fauna in different areas. The high population growth rate of many species will ensure changes in pest distribution.
3. If the rise in winter temperature takes place, the duration of hibernation of pests may decrease, thus increasing their activity.
4. Extension of geographical range of insect pests,
5. Increased over wintering and rapid population growth,
6. Changes in insect - host plant interactions
7. Increased risk of invasion by migrant pests,
8. Impact on arthropod diversity and extinction of species,
9. Changes in synchrony between insect pests and their crop hosts,

10. Reduced effectiveness of crop protection technologies
11. Insect growth, development, reproduction, survival rate and diapauses.

Table 3. Species diversity among different groups of organisms

Organisms	Number of species
Virus, algae , protozoa	80,000
Bacteria	4,000
Fungi	72,000
Plants	2,70,000
Animals invertebrate (Insects)	13,60,000
Animals	48,500
Total	**18,34,500**

Speciation takes between 100 and 1,000,000 years, providing between 10 and 10,000 new species per annum (Table 3). Nearly 99.9% of all species that ever existed have become extinct. The relative abundance of different insect species may change rapidly due to climate change, and the species unable to withstand the stresses may be lost in the near future. The current extinction rates are 100 to 1000 times greater than what has happened earlier, and nearly 45 to 275 species are becoming extinct every day.

Climate change on crop pest interaction

Crops and plants used as food by human beings are damaged by over 10,000 species of insects, and cause an estimated annual loss of 13.6 % globally and 23.3% in India. In India, the average annual losses have been estimated to be 17.5 % valued at US$ 17.28 billion in eight major filed crops (cotton, rice, maize, sugarcane, rapeseed - mustard, groundnut, pulses, coarse cereals and wheat). Host plant resistance to insects is one of the most environmental friendly components of pest management. However, climate change may alter the interaction between the insect pests ad their host plants. Insect –host plant interactions will change in response to the effects of CO_2 on nutritional quality and secondary metabolites of the host plants. Increased levels of CO_2 will enhance plant growth, but may also increase the damage caused by some phytophagous insects.

Sorghum varieties exhibiting resistance to sorghum midge, *Stenodiplosis sorghicola* (Coq) in India become susceptible to this pest under high humidity and moderate temperatures near the equator in Kenya. Increased CO_2 may also cause a slight decrease in nitrogen - based defenses (e.g. alkaloids) and a slight increase in carbon -based defenses (e.g., tannins). Acidification of water bodies by carbonic acid (due to high CO_2) will also affect the floral and faunal diversity. Lower foliar nitrogen content due to CO_2 causes an increase in food consumption by the herbivores up to 40 %, while unusually severe drought increases the damage by insect species such as spotted stem borer, *Chilopartellus* in sorghum.

Increased temperature and pest population

Temperature is the single most important factor influencing insect behaviour, distribution, development and reproduction. The mean temperature in India is projected to increase by 0.4 to 2^0 C during Kharif and to 1.1 to 4.5^0 C in Rabi by 2070. The swarms of locus produced in the Middle East usually fly eastward into Pakistan and India during summer season and they lay eggs during monsoon period. The swarms as a result of this breeding return during autumn to the area of winter rainfall, flying to all parts of India and influencing *kharif* crops.

The polycyclic diseases, such as blast, brown spot, bacterial leaf blight and rice tungro virus that invade the aerial parts of the plants, are subjected to constant interaction with weather. They easily attain epidemic proportions to cause heavy losses. Forecasting of aphids (*Lipaphiserysine* Kalt) on mustard crop, grown during winter season in northern part of India based on the movement of western disturbances, has been established. It was observed that there was a sharp increase in the population of aphids when the mean daily temperature ranged from 10 to 14^0C and relative humidity of 67-85 % cloudiness greater than 5 octas.

Some insects take several years to complete one life cycle- these insects (cicads, arctic moths) will tend to moderate temperature variability over the course of their life history. Some insects are closely tied to specific set of host crops. Temperature increases that cause farmers not to grow the host crop any longer would decrease the populations of insect pests specific to those crops. The same environmental factors that impact pest insects can impact their predators and parasitoids and entomopathogens resulting in changed status of biological control. At higher temperatures, aphids have been shown to be less responsive to the aphid alarm pheromone they release when under attack by insect predators and parasitoids - resulting in the potential for greater predation (Awmack *et al.*, 1997).

Insects and CO_2

Predicted increases in atmospheric CO_2 and global mean temperature are likely to influence insect –plant interactions. Plant traits important to insect herbivores, such as nitrogen content, may be directly affected by elevated CO_2, and temperature while insect herbivores are likely to be directly affected only by temperature. Flynn *et al.* (2006) stated that aphid populations did not change significantly under elevated CO_2, but tended to increase slightly. The changes in climate increases the likelihood of insect transport from regions to regions, as well. Migratory insects may arrive or the area in which they are able to overwinter may be expanded. Natural enemy and host insect populations may respond differently to changes in temperature

The growth and development of tobacco caterpillar, *Spodoptera litura* (Fabricius) (Noctuidae: Lepidoptera), reared on peanut (*Arachis hypogaea* L.) foliage grown under elevated CO_2 (550 ppm and 700 ppm) concentrations in open to chambers. Significantly lower leaf nitrogen, higher carbon, higher relative proportion of carbon to nitrogen and higher polyphenols content expressed in terms of tannic acid equivalents were observed in the peanut foliage grown under elevated ch as longer larval duration, higher larval weights, and increased consumption of peanut foliage by *S. litura* larvae under elevated CO_2 compared with ambient CO_2.

Parasitism could be reduced if host populations emerge and pass through vulnerable life stages before parasitoids emerge. Hosts may pass though vulnerable life stages more quickly at higher temperatures, reducing the window of opportunity for parasitoids. Increased carbon to nitrogen ratios in plant tissue resulting from increased CO_2 levels may slow insect development and increase the length of life stages vulnerable to attack by parasitoids (Coviella and Trumble, 1999).

Insects and their enemies

Climate change impacts not only the pest species but also their natural enemies viz., parasitoids, predators and pathogens which play crucial role in their population dynamics. Higher temperatures and CO_2 affect the tritrophic interaction chain involving host plant, host insect and its parasitoid or predator. Changes in foliar chemistry are likely to alter interactions between herbivorous insects and their natural enemies (Rao *et al.,* 2008).

Insects susceptibility to parasitoids, predators and pathogens can be altered due to these changes in host plant quality by affecting periods of exposure, physiological defense mechanisms and behaviour or by affecting the natural enemies themselves. Variation in the quality and composition of food consumed, which is likely to result in differences in the chemical composition of the herbivore can affect its susceptibility to predator attack (Price *et al.,* 1980).

The published information on impact of elevated CO_2 on higher trophic levels is very little and certainly elevated CO_2 affects the survival of natural enemies. It was reported that parasitism by hymenopteran parasitoid, *Cotesia melanoscelus* significantly prolonged gypsy moth feeding on elevated CO_2 grown aspen, birch, maple and oak, development and reduced growth rates. The only significant effect of elevated CO_2 on parasitoid included an increase in mortality and decline in adult female size.

Insect pollinators

Maintaining and increasing yields in horticultural crops, seeds and pastures through better conservation and management of pollinators is critically important to health , nutrition, food security and better farm incomes for poor farmers. The dependence of ecosystems on animal pollinators is even stronger in the tropics than the global average; less than 3% of all tropical lowland plants rely on wind for pollination. Climate change will exert great impacts on global ecosystems. Climate change may be a further threat to pollination services. Data on the impacts of climate change on crop pollination is still limited and no investigation has yet addressed this issue. In a recent review, the consequences of temperature induced changes in plant pollinator interactions were discussed. They found that timing of both plant flowering and pollinator activity seem to be strongly affected by temperature. Insects and plants may react differently to changed temperature, creating temporal and spatial mismatches with severe demographic consequences for the species involved Mismatches may affect plant by reduced insect visitation and pollen deposition, while pollinators experience reduced food availability. The effect of increasing temperatures on a highly resolved plant pollinator network. They found that shifts in phenology reduced the floral resources available for 17 to 50 percent of the pollinator species. A temporal mismatch can be detrimental to both plants and pollinators. However, the negative effect of this, there is an urgent need to investigate the thermal tolerance of important crop pollinators and differences in thermal tolerance among *Apis* species and sub –species. Some of these are better adapted to warmer climates and may therefore may move into new areas where they can function as plant pollinators under future climate conditions. The effect of climate change on pollinators depend upon their thermal tolerance and plasticity to temperature.

Climate change on termite

Precipitation changes will also have an effect. Termites, being soft - bodied are prone to water loss and death by desiccation in soils with less than about 8 to 10 per cent moisture by weight. Most termites die off during the winter in areas where the ground freezes: most species cannot survive sustained temperatures below 9 degrees Celsius, and even fewer can survive temperatures below 1 degree Celcius. Thus the increase in the winter temperature in the North will affect termite populations largely because as winter temperatures warm, termites may be better able to survive winters in more northern climates.

Higher summer temperatures in the North might allow any termite populations already present to increase in size. In Southern states, a newly formed termite colony starts with two reproductive individuals and then builds for several years until it reaches a certain size. At that point, winged, or alate, termites capable of reproduction begin to appear. In the spring, thousands of these alates from many colonies take flight (sometimes called a swarm). Adults pair up and start new colonies, reinitiating the cycle. Enhanced summer rainfall and drought conditions on soil dwelling. *Agrioteslineatus* (wireworms) promote rapid increase in the population of wireworms in upper soil.

❑❑❑

Chapter 16

Methane Emission from Rice Field

Methane is a colourless, tasteless, odourless saturated carbon gas. Methane may be produced by transmethylation of acetic acid and by the reduction of carbon dioxide by hydrogen ions, fatty acids and alcohols. Methane may be referred as marshy gas. Paddy fields account for around 20% of human related emissions of methane, a potent greenhouse gas. Draining paddy fields in the middle of the rice growing season saves water, stopped most of the methane release from the field. Methane is relatively an active trace gas with about 25 to 30 times more infrared sorbing capability per molecule than carbon dioxide. The atmospheric concentration of methane has increased from 0.7 to 1.72 ppm. Atmospheric methane concentration is currently increasing at the rate of 0.8% per year and can double by 2100 AD. Methane contributes about 15 to 20 percent of global warming due to its atmospheric residence of 8 to 10 years. It requires about 10 percent reduction in anthropogenic emission of methane to stabilize its atmospheric concentration at the present level. Whereas for other gases, much higher folds of reduction would be required.

Methane is one of the two end-products of organic matter decomposition by strictly anaerobic microorganisms. Predominantly anaerobic flooded soils, an excellent medium for the growth and activities of anaerobic methanogenic bacteria, can serve as the major source of atmospheric methane. Rice fields, under intensive use of nitrogen fertilizers in modern agriculture, can also contribute significantly to atmospheric nitrous oxide. Particularly under rainfed rice, where intermittent flooding and drying cycles are common and widespread especially under erratic rainfall as in eastern regions of India, nitrous oxide emission can be high. The total global area of rice is about $147.5X10^{10}m^2$ and it is estimated to contribute about 25.4 to

54 Tg of methane annually. India, with 42.2 percent of the total global harvested area and 19.5 percent of the rice grain yield, was reported to contribute about 10.5 percent of the total global emission of methane.

Methane formation in paddy field

As in a natural wetland, flooding a rice field cuts off the oxygen supply from the atmosphere to the soil, which results in anaerobic fermentation of soil organic matter. Methane is a major end product of anaerobic fermentation. It is released from submerged soils to the atmosphere by diffusion and through roots and stems of rice plants. Methane is produced as the terminal step of the anaerobic breakdown of organic matter in wetland rice soils. Methane is exclusively produced by methanogenic bacteria that can metabolize only in the strict absence of free oxygen and at redox potentials of less than -150 mV. Most methanogens are neutrophilic, with an optimal pH of 6-8. Methanogens rely on a plethora of other microorganisms to provide them with the few substrates they can catabolize: hydrogen, carbon dioxide, acetate, methanol, methylamines, and methy sulphides. In wetland rice soils, methane is largely produced by transmethylation of acetic acids and, to some extent, by the reduction of carbon dioxide. When the paddy soil is submerged, some part of organic carbon in soil is decomposed to carbon dioxide with the reduction of the accepter. After iron oxide is reduced, the organic carbon in soil is an aerobically decomposed to methane.

The rate and pattern of organic matter addition and decomposition determine the rate and pattern of methane formation. Methane production generally increases during the cropping season, although the population density of methanogens remain fairly stable. Easily degradable crop residues, fallow weeds, and soil organic matter are the major source for initial methane production. At later growth stages of rice, root exudates, decaying roots, and aquatic biomass seem to be more important. Methane production is enhanced in the rooted soil zones.

Factors affecting methane emissions

Stage of the crop

Methane emission gradually increases during the vegetative phase correlating with increasing plant biomass, and peaks near panicle differentiation, a period of rapid root development. Emission remains relatively constant during the reproductive stage, but may decrease during late grain filling because of root degradation. Prior to the end of the season, a second emission peak may be observed. This late-season increase in emission can be attributed to an increase in soil C substrate due to accelerating leaf and root senescence. The addition of readily degradable C, such as rice straw, before planting results in an increase in the early season methane emission. The decomposition of the straw may also result in an enhanced or additional early season emission maximum.

Photosynthetic activity

In irrigated tropical rice with double cropping, both methane emission and rice grain yields are consistently higher in the dry season than in the wet season. One interpretation of this finding is that higher photosynthetic rates during the sunnier days of the dry season lead to larger amounts of C available to methanogenic bacteria and, consequently, greater production and emission rates of methane. Seasonal rates of methane emission and rice grain yield have been positively correlated with accumulated solar radiation. A 1% increase in accumulative solar radiation is accompanied by a 1.1% increase in methane emission and a 1% increase in rice grain yield.

Organic Amendments

Organic amendments of flooded rice increase both methane production and emission to the atmosphere. The amount of methane emitted as a result of organic soil amendments depends greatly on the amount and condition of readily available decomposable C contained in the treatment. Methane emission with incorporated rice straw has been found to be higher than that from either compost or mineral-fertilized plots. The effect of incorporating green manure is even more dramatic.

Inorganic Fertilization

The use of ammonium sulphate fertilizer may reduce methane emission through the competition by sulphate-reducing bacteria for hydrogen or the presence of hydrogen sulphide toxicity.

Water Management

Methane emission rates are highly sensitive to soil aeration controlled through water management. Periodic drainage of irrigated rice which significantly decreases methane emissions, may be the most effective method of mitigating them. In addition, multiple short periods of drainage (2-3 days) approximately every 3 weeks during the growing season reduced methane emissions to an insignificant amount.

Soils

Rice soils prone to methane production belong mainly to the orders of Entisols, Inceptisols, Alfisons, Vertisols, and Mollisols. Oxisols most of the Ultisols, and some of the Aridisols, Entisols, and Inceptisols are less favorable to methane production when flooded. After the flooding of calcareous or alkaline soils (high pH), methane production may occur almost immediately, whereas emission may take weeks to develop with acid soils and may not occur at all in highly acid soils. In general, sandy soils high in organic C produce more methane than clay soils with similar C content.

Rice Cultivars

Methane emission from eight different cultivars grown under comparable conditions near New Delhi, India differed by as much as one order of magnitude. Fields planted with the Italian rice cultivar Lido showed methane emissions 24-31% lower than fields planted with the cultivar Roma. The difference in methane emissions was attributed to a significantly lower gas transport capacity in the Lido cultivar. Cultivars exhibiting higher seasonal methane emissions also exhibited higher soil acetate concentrations, suggesting that intervarietal differences in methane emissions may be the result of different soil organic substrate levels and, hence, different rates of methane production.

Mitigation of Methane Emission

There are efforts worldwide to mitigate the emissions of methane from rice. Intermittent flooding and drainage is one the practices recommended for reducing methane emissions especially from irrigated rice fields. But, in rainfed rice culture in the eastern and north-eastern regions of India, drainage of water from rice fields is virtually impossible due to high water table in these areas. Some cultural practices used in rice culture have the additional advantage of mitigating the emission of methane from rice fields. Moreover, intermittent flooding and drainage of rice fields may retard methane emission but, on the other hand, promote the emission of nitrous oxide through enhanced nitrification-denitrification especially from rice fields. For instance, application of single super phosphate to flooded rice fields effected a distinct inhibition of emission of methane from fields and its production under laboratory condition. Likewise, application of Hexa Chlorocycle Hexane (HCH) to flooded fields or its individual isomers to rice soil inhibited the production and emission of methane. In addition to the influence on methane producers, HCH also retarded the methane oxidizing activities in flooded soils. Thus, management practices used in rice culture, such as intermittent flooding and drainage and application of certain agrochemicals would help in mitigating the emission of methane from rice fields.

The possible ways to minimize methane emission in rice include

1. Avoiding deep ploughing cum puddling to minimize the depth of reduced environment
2. Pulling out seedlings at appropriate age than leaving them for longer period in nursery for want of labour etc. for transplanting.
3. Use of well decomposed or composted organic amendments.
4. Avoiding basal application of nitrogen if green leaf manure or farm yard manure is applied basally.
5. Leaving sufficient gap between green leaf trampling and planting so as to oxidize methane formed in the soil in situ.
6. Using sulphate forms of inputs or sulphur coated inputs.
7. Application of gypsum.
8. Applying optimum doses of inputs based on soil test rather than blanket use.
9. Choosing varieties with lesser emission rates.
10. Adopting an ideal spacing to get maximum yield with less emission rates.
11. Adopting stubble management.

12. Water control is one of the most important factors in rice production. Mid-season drainage and alternative flooding/ drying management reduced methane emissions by 23 to 59%, while yield increased by 16%. Water management would be the most promising mitigation option.
13. Application of organic manure is a common practice to maintain soil fertility, but it increases methane emission from rice fields. The effect of methane emission may be reduced by composting manure and rice straw, rather than applying raw fresh materials.
14. Rotation application of organic amendment and mineral fertilizer can reduce methane emission of 139 to 279 $mg/m^2/day$.

❑❑❑

Chapter 17

Effect of Ozone and UV-B on Crop Growth and Productivity

Ozone

There are number of recent reviews of the mechanism of action of O_3 on plants and of crop response to O_3. The response of plants to O_3 may be viewed as the culmination of a sequence of physical, biochemical and physiological events. The O_3 diffuses from the air into the leaf through stomata and these exert control on O_3 uptake. Plants are able to detoxify O_3 or its metabolites and can repair or compensate for O_3 impacts, so visible signs of ozone injury may not occur if the rate of O_3 uptake is sufficiently small. When resources have to be devoted to repair or compensation mechanisms or when O_3 uptake is too large for full repair to be achieved, growth and yield may be reduced in the absence of visible injury by O_3.

The principal modes of action of O_3 on plants are through injury to proteins and membranes, reduction in photosynthesis (i.e., carbon gained), changes in allocation of carbohydrates and acceleration of senescence. The main result of O_3 exposure is therefore, a reduction in the capacity of the plant to accumulate photosynthate through loss in photosynthetic capacity and increased foliage senescence.

Crop sensitivity to Ozone

1. Over 20% of the crop production land in Europe is at risk for yield losses of 5% or more due to O_3 pollution.
2. Ozone unpolluted region has ozone concentration of 20 - 50 ppb (60 ppb).

3. Ozone polluted region has ozone concentration of 400 ppb.
4. Low concentration of ozone i.e., 30 ppb is toxic to plants.
5. Crop responses to ozone are given below (Table 4).

Table 4. Crop responses to ozone

Sensitive	Moderately sensitive	Tolerant
Wheat	Sugarbeet	Barley
Watermelon	Potato	Plum
Pulses	Oilseed rape	Strawberry
Turnip	Tobacco	
Tomato	Maize	
Onion	Grapes	
Lettuce	Broccoli	
Soybean		

Effect of UV-B on plants

- Bronze or brown spots on the leaf surface and chlorosis, necrosis, and desiccation of the leaves.
- Continuous exposure of leaves to UV-B leads to involuted or cup-shaped leaves and finally leaves dry up.
- Increase in leaf thickness.
- In cotton, decrease in leaf thickness along with an increase in number of palisade layers was observed.
- In susceptible rice cultivars, leaf cell structure was damaged when they were exposed to UV-B radiation of 15-16 kJm^{-2} per day.
- Maize leaves when exposed to UV-B radiation of 9 kJm^{-2} per day, collapse of the adaxial epidermis in the distal region of the leaf and decrease in fractional volume of chloroplasts were observed.
- Reduces mainstem and branch elongation rates.
- Decreased plant height, shorter internodes rather than fewer nodes and fewer cells rather than reduced cell length.
- Leaf area was less and smaller and lesser number of leaves were observed.

- Reduction in leaf cell size, number, cell division and cell expansion.
- Exposure to enhanced UV-B radiation leads to decrease in chlorophyll content -.
- Chlorophyll reduction was upto 10 to 70%.
- Chlorophyll reduction is higher among the dicot species (10-78%) than in monocot species (0-33%).
- Ultrastructural damage to chloroplasts and changes in photosynthetic pigments may reduction in photosynthesis.
- Crop responses to UVB radiation are furnished below (Table 5).

Table 5. Crop responses to UV B

Sensitive	Moderately sensitive	Resistant
Barley	Rice	Corn and millet
Oats	Rye	Egg plant
Sweet corn	Sorghum	Orange
Soybean	Bean	Artichoke
Pea	Pepper	Cotton
Cowpea	Potato	Wheat
Tomato		Sunflower
Cucumber		Asparagus
Squash		Onion
Okra		Tobacco
Pumpkin		Radish
Melon		Cabbage
Red raspberry		
Blueberry		
Cauliflower		
Broccoli		
Sugarcane		

Interactive effect of Ozone and UV-B radiation on crops

Krupa and Kickert (1989) assessed potential risks on a geographical basis, using distributions of major crops, tropospheric ozone concentrations and UV-B irradiance. They suggested that interactions might involve episodic exposure to the two stresses, so that peaks of O_3would coincide with lower UV-B irradiance at the surface is reduced, because the O_3 absorbs some UV (Table 6).

Table 6. Interactive responses to crops to ozone and UV B radiation

Crop type	Crop	Sensitivity to UV-B	Sensitivity to O_3
Fibre crops	Cotton	Tolerant	Sensitive
C_4 grain plants	Sorghum	Sensitive	Intermediate
Legume seeds	Pea	Sensitive	Sensitive
Legume seeds	Beans	Sensitive	Sensitive/inter.
C_3 grain crops	Barley	Sensitive	Tolerant
Roots & tubers	Potato	Sensitive/Tolerant	Sensitive
Legume seeds	Soybean	Sensitive	Tolerant
C_4 grain crops	Corn	Tolerant	Sensitive
C_3 grain crops	Oats	Sensitive	Sensitive
C_3 grain crops	Wheat	Tolerant	Intermediate
Fruit crops	Cucumber	Sensitive	Intermediate
C_3 grain crops	Rice	Sensitive	Intermediate
Fruit crops	Tomato	Sensitive	Sensitive/inter.
Leaf crops	Cabbage	Tolerant	Intermediate
Flower crops	Chrysan-themum	Tolerant	Intermediate

The measured solar spectral irradiance (I) at the ground near Nottingham, England, on two days (12 and 18 July 1990) when the tropospheric O_3 concentration was 51 and 84 nmol/mol, respectively and stratospheric O_3 was constant. The results revealed that Irradiance (I) was about 20 to 40% greater in the UV-B waveband (280-320nm) on the low O_3 day than on the high O_3 day. However Irradiance (I) at longer wavelengths was also greater (by about 20-25%), and this increase is most probably a consequence of less aerosol (dust) being present on the low O_3 day. Since this aerosol effect would also apply in the UV-B waveband, it seems likely that the UV-B increase directly attributable to reduced tropospheric O_3 was no more than 10% to

15%. Variations in aerosol from day to day seem likely to be more important than variations in tropospheric O_3 in modulating the intensity of UV-B at the ground, but further observations would be valuable.

The ozone treatments resulted in accelerated senescence and reduced yield, in agreement with many other published studies. In contrast, the UV treatments, even at this relatively large UV irradiance, did not induce any significant changes in photosynthesis or yield and there were no UV/O_3 interactions.

Fiscus *et al.* (1994) attempted to reconcile the lack of response to UV reported in the Raleigh experiments with other reports which indicate that increased UV-B causes physiological disfunction and reductions in crop yields. They concluded that three factors make the conclusions of other field and glasshouse studies, at best, hard to interpret and possibly misleading: a frequent failure to monitor UV-B adequately ; a tendency to underestimate UV exposures when relying on model calculations which do not allow sufficiently for seasonal and daily weather changes (and hence a tendency towards unrealistically large exposures).

□□□

Chapter 18

Climate Change on Pulses

Climatic conditions vary tremendously within the northern Great Plains, including long, cold winters; short, warm summers; large diurnal ranges in temperature; frequent strong winds; highly variable and unpredictable precipi-tation. The precipitation uncertainty and temperature extremes are serious risks to agriculture on the Plains. However, pulse crops provide producers with opportunities to diversify cropping systems and assist with managing the risk of unpredictable weather and market patterns. Pulse crops also complement cropping systems such as no-till or direct-seeding. The adoption of pulse crops has enabled producers to reduce summer fallow and increase cropping intensity because of improved soil conservation and increased soil water availability. The diversification and intensification of no-till cropping systems have significantly contributed to increase environmental and economic sus-tainability. Pulse crops increase market diversification since their prices respond some-what independently of cereal grain markets. Pulse crops increase production diversification due to differential responses to growing season, rainfall and temperature patterns. The inclusion of pulses in crop rotations often increases the efficiency of cereal crop production.

The area sown to pulse crops in the northern Great Plains has increased steadily in the last two decades. The most notable area for increased production has been in semiarid regions of the Canadian Prairies, where dry pea, chickpea, and lentil are used to extend the traditional wheat-fallow crop rotations, especially in no-till management systems.

Pulses are grown on 22-23 million hectares of area with an annual production of 13-15 mt. India accounts for 33% of the world area and 22% of the world production of pulses. About 90% of the global pigeonpea, 65% of chickpea and 37% of lentil area falls in India, corresponding to 93%, 68% and 32% of the global production, respectively. Due to stagnant production, the net availability of pulses has come down from 60 gm/day/person in 1951 to 31 gm/day/person (ICMR recommends 65 gm/day/capita) in 2008.

Response of red gram to elevated CO_2

The variety ICPL 88039 was evaluated at ambient (370 ppm) and twice the ambient (700 ppm) concentrations of CO_2 in OTCs. Significant positive enhanced response for total biomass, fodder yield, grain yield, number of pods and seeds per plant, test weight and HI was observed at elevated CO_2. Under elevated CO_2, the total biomass recorded an improvement of 91.3%, grain yield 150.1%, fodder yield 67.1%. The major contributing components for improved grain yield were number of pods, number of seeds and test weight which recorded an increase of 97.9%, 119.5% and 7.2%, respectively. Increase in HI was due to its improved pod set and seed yield. Total biomass (grams per plant) improved from 61.1 g/plant ambient to 116.9 g/plant under elevated CO_2 mainly due to, Increased photosynthesis in all C_3 plants.

Responses of blackgram under elevated CO_2 and moisture stress

Interactive eûect of the elevated CO_2 and moisture stress on germination and initial growth responses of blackgarm and four open top chambers (OTCs) were used for diûerent conditions viz., irrigated with ambient CO_2(365 ppm), irrigated with elevated CO_2(600 ppm), moisture stress with ambient CO_2 and moisture stress with elevated CO_2. The percentage of germination, germination speed, emergence index, and vigor index were measured on the 5th and 6th DAS. It was found that percentage of germination, speed of germination, emergence index and vigor index were positively increased with the elevated CO_2. The increment of the above-mentioned parameters was higher under moisture stress conditions than under irrigated conditions.

A study was conducted to quantify the interactive effects of elevated temperature and CO_2 on photosynthesis, pollen production and viability, seed set, seed number, seeds per pod, seed size and seed yield and also to determine if deleterious effects of high temperature on reproductive processes and yield could be compensated by enhanced photosynthesis at elevated CO_2 levels (Prasad *et al.*, 2002).

They observed that, the percentage of increase in root length was higher under elevated than ambient CO_2 both under irrigated and stress conditions, root volume was also increased due to more lateral roots and root hair formation under elevated CO_2 condition and proportioning of assimilates was greater to the roots than to the shoots under stress. No effect of temperature or CO_2 on pod settings between 28-37° was observed and pod set was averaged about 67%. However, exposure to 40° reduce the pod sets to 15% and 10 % in ambient CO_2 and elevated CO_2. As the temperature increased from 28-40^0 C, seed set was reduced by 6% per degree rise in temperature at both aCO_2 and eCO_2. The ceiling temperature were 40° and 38° for aCO_2 and eCO_2 (Prasad *et al.*, 2002). At temperature >32°C, the pollen number was reduced to 250 per flower and no pollen was produced at 40°C. When the ceiling temperature of aCO_2 and eCO_2 is 37°, elevated CO_2 increased the seed number from 9.8 to 12.4 per plant and seed yield from 4.6 to 5.6 g/plant.

When the leaf temperature grown at eCO_2 level was about 1.5 degrees higher than those at aCO_2 level, it was observed that

- No beneficial interaction of elevated CO_2 with higher temperature in bean.
- At both levels of CO_2, higher temperature resulted in significant yield loss.
- The beneficial effects of CO_2 on photosynthesis and growth rate were overwhelmed by negative effects of higher temperature during reproductive growth.
- High temperature stress during reproductive development decreased seed yield of chickpea by decreasing the number of seeds per plant and weight per seed primarily and the harvest index secondarily.
- High temperature stress imposed during pod development caused more yield loss than the stress during early flowering.

❒❒❒

Chapter 19

Climate Change on Cotton

Cotton is the most widely produced natural fibre in the world and represents about 46 per cent of the global textile market. Cotton seed is a by-product of the valuable cotton fibre, and makes up about 15 per cent of the total financial returns to farmers. Cotton seed is a valued raw material for food oils for human consumption and high protein feed for livestock. Cotton seed oil is one of the world's most popular vegetable oils.

Cotton belongs to the C_3 plant, which releases CO_2 during photorespiration. High external input and over use of N fertilizers lead to more emission of nitrous oxide. Global greenhouse gas emissions have grown since pre-industrial times with an increase of 70% between 1970 and 2004 (IPCC, 2007). Atmospheric carbon dioxide levels are rising rapidly. In 2005, the globally averaged concentration of carbon dioxide in the atmosphere was about 379 ppm, one-third higher than it was before the industrial revolution began in the late 18th century. The earth temperature has increased by 0.74°C during the last century (1906 to 2005) due to increase in greenhouse gases through anthropogenic emissions as reported by IPCC. Thus, the increase in temperature is likely to be 1.8–4.0°C by the turn of 21st century resulting in anticipated greater instability in food, feed and fibre production. Increase in temperature can reduce crop duration, change pest populations, hasten mineralization in soils and increase evapo-transpiration. It is reported that 40 and 50% less biomass is anticipated in cotton at 20/10°C and 40/30°C, respectively, with optimum temperature of 30/20°C.

Cotton species of *G. barbadense* showed more sensitive than *G. hirsutum*. *G. arboreum* is suitable for low and erratic rainfall with drought situations. In saline environment, *G. herbaceum* showed better adaptability. The available drought tolerance *hirsutum* genotypes, like 'LRA 5166', 'KC 2' and 'AKH 081' may show better adaptation. The risk and uncertainty imposed by climate change could be managed by adoption of location-specific intercropping and multi-tier cropping system. In situ soil moisture conservation techniques include contour bunding, graded, narrow or broad ridges or beds separated by furrows, ridges and furrow, opening of furrow after every rows of cotton, black polythene mulch (25 microns), and spreads of crop residue were found to be promising. Sensitivity of cotton to greenhouse gases are presented (Table 7).

Table 7. Sensitivity of cotton to green house gases

Crop type	Crop	Sensitivity to UV-B	Sensitivity to O_3	Sensitivity to SO_2
Fibre crops	Cotton	Tolerant	Sensitive	Sensitive

Table 8. Temperature for different growth stages of cotton

Growth stage (day/month)	T_0 (°C)	T_1 (°C)	T_2 (°C)
Seeding stage	11	26	35
Sprouting stage	15	26	35
Flower bud stage	19	28	35
Flowering stage	15	26	35
Boll opening	11	26	32

(Where T is average temperature, T_0 T_1 and T_2 respectively refer to the bottom, the optimum and the upper temperature).

Important effects

Temperature for different growth stages are given in Table 8. Higher temperature causes premature leaf senescence, decreased light interception, decreased chlorophyll content and photosynthesis, reduction in assimilate availability, alterations in assimilate partitioning, flowering and pollen tube extension.

Germination

- Delayed germination in the seeds. Decrease in germination ranged from 55 to 100% - plants.

Pollen germination, flower initiation and reproduction

- Reduces pollen germination and / or pollen tube extension.
- Reduced flowering. 21 days exposure to 70 µl l^{-1} O_3 for 6 h/day delayed and reduced flowering.
- 5 days @ 6 h/d exposure to 100 µl l^{-1} O_3 increased flower abortion.

Biochemical processes

- Reduction in CO_2 fixation by ribulose-bis-phosphate carboxylase.
- Inhibition of CO_2 assimilation, stomatal opening.
- Reduced photosynthesis - Oxidation of –SH groups of Rubisco.
- Reduced assimilation and increased respiratory loss of CO_2
- Reduced leaf area - Less light interception - Less assimilates.
- Nitrogen metabolic cycle - Changes in amino acid and protein content.

Plant growth

- Inhibited photosynthesis, carbon (sugar) production, altered carbon allocation to roots and stems, uptake of minerals (N, P, K and S).
- Advanced senescence and accelerated catalysis - Rapid loss of protein and chlorophyll.
- O_3 impairs phloem loading and assimilate partioning.
- Increased shoot sink - Repair of leaves - Carbon limitation to roots.
- Higher shoot/root biomass ratio, lower harvest index (reduced grain filling period), altered leaf chemistry.
- Decrease root exudates and poor water dissolved organic carbon (DOC).
- Reduced leaf N content - affect quality of litter.
- Changes in 'N' cycle - affects nitrification and denitrification.

Climate change on cotton growth and development

Green house gases - Carbon dioxide (CO_2)

At the leaf level, the two most well-known responses to elevated CO_2 are an increase in the rate of net photosynthesis, P_N, and a decrease in stomatal conductance, g_s. The increase in P is much larger in C_3 species (50 to 100% when CO_2 concentration doubles) than in C_4 species (10%), but a substantial decrease in g (30 to 40%) is observed in both types of species. There is some evidence of photosynthetic acclimation to elevated CO_2, so that increased rates of P_N do not persist at the leaf level in long-term studies with some species, but there is no evidence of acclimation in g. It is not clear whether increased CO_2 also indirectly affects plant growth, e.g., by changing rates of leaf initiation, expansion or longevity, but such effects could have a large influence on crop productivity.

Much of our understanding of effects of CO_2 on plants has been gained from studies with individual leaves. There have been far fewer studies of long-term effects of elevated CO_2 on canopy-scale photosynthesis and transpiration. Most of those reported have been in closed- or open-top chambers, though a few reports of field-scale exposures (FACE) are now becoming available. Drake and Leadley (1991) summarized the data available at that time and concluded that: (1) canopy photosynthesis increases in elevated CO_2 when there is a sink available for the carbon; (2) the relative effect of CO_2 is greatest at highest temperatures; and (3) elevated CO_2 alters many interacting factors, such as canopy architecture and partitioning of assimilates, that mediate gas exchange of canopies and ecosystems.

As a consequence of increased net photosynthesis, dry matter production and yield are substantially increased by elevated CO_2, provided there is adequate water, nutrients and pest control, yields of C_3 and C_4 crops growing in about 700 μ mol/mol CO_2 would be about 30 to 40% and 9%, respectively, greater than present yields (350 μ mol/mol CO_2) if all other climatic factors were unchanged. Although harvest index (seed or pod yield per plant) was not altered by CO_2, yield increases were consistent with other reports for unstressed C_3 and C_4 crops.

However, in much of the world, ideal crop growing conditions, with adequate water and nutrition, are wishful thinking, and the influence of elevated CO_2 on aspects such as water-use may be much more relevant. Reviews by Morison (1985) and Eamus (1991) make it clear that, on a leaf area basis, elevated CO_2 improves the water-use efficiency (WUE - the ratio of weight of dry matter produced to weight of water transpired) of plants. On a whole-plant basis, the increase in leaf area resulting from enhanced photosynthesis may cause the amount of water-used per plant to remain close to that of plants growing at current levels of CO_2, although the WUE as defined here increases.

CO_2 and temperature

The atmospheric environment of the future is likely to include correlated increases of atmospheric CO_2 concentration and temperature. Long (1991) reviewed the mechanisms by which temperature and CO_2 affect photosynthesis in C_3 species, and developed models of the response of leaf canopy carbon exchange to changes in these variables. The interaction of carbon dioxide concentration and temperature causes the temperature optimum of the light-saturated rate of CO_2 uptake to increase as CO_2 increases. Without this interaction, an increase in global temperature of 2.5°C would increase the frequency of temperatures in temperate regions that are supra-optimal for light-saturated photosynthesis. However, if warming on this scale was coincident with a CO_2 concentration increase to above 500 μ mol/mol, the temperature optimum would increase, and this inhibition would not occur. Long (1991) went on to develop a simple model of canopy CO_2 uptake in response to temperature, taking into account the substantial fraction of the canopy leaf area that would be shaded and therefore not photosynthesizing at the saturated rate.

Interactions between temperature and CO_2 also need to account for effect of temperature on development. Squire and Unsworth (1990) pointed out that determinate crops with discrete elements to their life cycle develop faster in higher temperatures, and so the stage of seed filling is shortened, limiting the benefits of elevated CO_2. When CO_2 concentration doubled, but daily weather data from a typical year were used, the potential grain yield was 27% larger than in a control run at 340 μ mol/mol CO_2. The date of maturity was unchanged. When daily temperatures were increased by 3°C and CO_2 was doubled, more rapid development of the crop shortened the growing season, and the potential grain yield was only 15% larger than in the control, but maturity occurred 30 days earlier. Climatic warming could disrupt this sequence, for example, by providing insufficient winter chilling to synchronize spring bud break, or by advancing ovule development and leading to poor fruit set in spring (Cannell *et al.*, 1989).

CO_2 and water availability

The influence of CO_2 on the water-use of crops may well prove to be the most important benefit of increased CO_2 concentrations for agriculture. The effect at a range of scales, pointing out that there are now many studies, including annual cereal crops, legumes and woody perennials, which show that, contrary to earlier assumptions, although the absolute amount of dry matter produced decreases at all CO_2 concentrations as water availability is reduced, the *relative* effect of increasing CO_2 on plant growth increases with decreasing water supply. There are at least two mechanisms responsible for this relative enhancement: (1) the reduction in stomatal conductance, with a consequent beneficial effect on leaf water potential and hence on leaf expansion; (2) greater allocation of carbon to roots in elevated CO_2, and hence improved potential for exploring the soil volume for water and nutrients. In contrast, the increased production of leaf area in elevated CO_2 acts to increase the water-use per unit ground area of crops.

It is less clear whether increasing CO_2 will decrease water-use of crop canopies in the field over periods of days or months. This is not only because of the various mechanisms described above, but also because soil evaporation must be accounted for, and feedbacks between leaf temperature, atmospheric humidity and canopy evaporation. Consequently, the very rough estimate that a 33% increase in dry matter production combined with a 33% decrease in transpiration in response to doubling CO_2 concentration would increase water-use efficiency by a factor of 2 is unlikely to apply. While the total water-use of the canopy was unaffected by growth in elevated CO_2. WUE was increased by 25% on a canopy basis or 87% on a leaf area basis.

CO_2 and ozone

The interactive responses of crops to CO_2 and O_3 describing experimental work was not available but the effects of ozone alone available. Allen (1990) made an estimate of the possible scale by which CO_2 might influence responses of crops to O_3 to yield (Heagle *et al.*,

1983). According to these models, doubling CO_2 concentration would reduce g by about 30% and thus stomatal uptake of O_3 by the crop. Allen (1990) estimated that this would result in a yield decrease of about 15% compared with yield in ambient air and also no interactions between the two gases.

In clean air, the typical effects of elevated CO_2 on photosynthesis (increased) and stomatal conductance (decreased) were observed; the extra carbon assimilation stimulated root growth rate by 43%, but there was no significant effect on shoot growth or leaf area. At ambient CO_2, the high O_3 treatment depressed photosynthesis by 26% (compared to the low treatment) and induced a slight reduction in g, with the net result that water-use efficiency declined. The reduction in carbon uptake was reflected in reduced growth, with roots more affected than shoots. Interactive effects of ozone and CO_2 were generally consistent with a reduction in ozone-induced responses in elevated CO_2.

Long-term exposure to O_3 resulted in a decrease in the light-saturated rate of CO_2 assimilation, partial stomatal closure, and accumulation of water soluble carbohydrate and starch in leaves. These resulted in decreased growth, with root growth more severely affected than shoot growth. In plants exposed to elevated CO_2 and O_3, effects on stomatal conductance were less than additive, and CO_2 enhancement effects on photosynthesis and growth were reduced by O_3. The O_3 exposure necessary to induce stomatal closure is generally sufficient to cause visible leaf injury, and this disbenefit is likely to be far more damaging than reduced CO_2 uptake through stomata. More field-based studies are necessary to assess fully the effects of the two gases in combination.

CO_2 and UV-B radiation

Many studies of effects of UV-B radiation on crops are seriously flawed because UV exposures were either inadequately specified and/or were unrealistically large. Effects of UV-B radiation on plants, pointed out that, since the spectrum of UV from artificial light sources differs from that in the solar spectrum, and since many photobiological processes are strongly wavelength dependent. Unfortunately, action spectra for many responses remain unknown. In addition, many plant responses to UV-B are larger when plants are growing at low light levels typical of growth chambers and some glasshouses than they are in full sunshine, probably because natural protective pigments are inadequately synthesized and repair processes are restricted in low light. For this reason, field-based studies of crops are most relevant for estimating responses of productivity and yield to UV-B. Even so, most field studies have used continuous UV-B exposures corresponding to 15 to 25% reductions in stratospheric O_3 i.e. roughly 30 to 50% increases in UV-B above present values. Such increases are much larger than are likely to occur as long-term means in agricultural regions under any likely depletion of stratospheric ozone in the next 30 years.

Effects of UV-B on yield of crops growing in the field have produced very variable results. Yield was reduced by 20 to 25% at the high UV treatment; under any realistic scenario of increasing UV-B radiation, crop yields are unlikely to be altered to a significant extent. Effects of the combination of elevated CO_2 and UV-B radiation on crop growth and yield have been addressed in very few studies. Teramura *et al.* (1990) grew wheat, rice and soybean in a glasshouse experiment with the following treatments in the factorial design: ambient CO_2 (350 (μ mol/mol), ambient UV, elevated CO_2 (650 μ mol/mol) and elevated UV-B (corresponding to a 10% decrease in stratospheric O_3 at the equator). Compared to the control, yield and total biomass increased significantly in elevated CO_2. However, when UV-B and CO_2 were increased simultaneously, no increase in either yield or total biomass was observed compared to the control. The incomplete experimental design limits interpretation of this study, but the results suggest that temperature and CO_2 are much more important influences on growth than any likely realistic increases in UV-B.

Ozone

The response of plants to O_3 may be viewed as the culmination of a sequence of physical, biochemical and physiological events. The O_3 diffuses from the air into the leaf through stomata, and these exert control on O_3 uptake. Plants are able to detoxify O_3 or its metabolites and can repair or compensate for O_3 impacts, so visible signs of ozone injury may not occur if the rate of O_3 uptake is sufficiently small. When resources have to be devoted to repair or compensation mechanisms, or when O_3 uptake is too large for full repair to be achieved, growth and yield may be reduced in the absence of visible injury by O_3. The principal modes of action of O_3 on plants are through injury to proteins and membranes, reduction in photosynthesis (i.e., carbon gained), changes in allocation of carbohydrate, and acceleration of senescence. The main result of O_3 exposure is therefore a reduction in the capacity of the plant to accumulate photosynthate through loss in photosynthetic capacity and increased foliage senescence.

Ozone and water availability

The availability of soil water is one of the strongest influences on crop growth and productivity. When available, irrigation is commonly used to alleviate water stress, but in conditions of large evaporative demand, even irrigated crops may experience water stress. The most immediate effects of water stress are on plant-water potential, and may lead to reductions in leaf expansion rates and altered partitioning of dry matter, with increased assimilate partitioning to roots. When stomatal closure occurs, there is a resulting reduction in transpiration and CO_2 uptake and an increase in leaf temperature.

Water stress and O_3 responses

There has long been a general belief that water stress reduces the magnitude of any adverse effects of O_3 i.e., leaf injury, growth and yield reductions. This belief was based on observations that stomatal closure, induced by drought, reduced the rate and quantity of O_3 absorbed by leaves (US EPA, 1986). However, more recent results show that the O_3/water availability interaction is complex, and probably depends on the magnitude and timing of the water stress. One difficulty in quantifying the interaction is the very large effect that water stress alone may have on yield, so that any influence of O_3 is sometimes masked.

Temple *et al.* (1985) studied the effects of O_3 and water stress on cotton growing in California. In this typically hot, dry season, the water-stressed plants wilted frequently, and yielded much less than the well-watered controls; nevertheless, when compared to yields of plants in a background O_3 concentration of 25 ç mol/mol, and expressed as a percentage, the estimated yield loss at a seasonal mean 0, concentration of 50 ç mol/mol was 7% in the well-watered plants and 2% in the water-stressed crop. Clearly the absolute yield loss from drought was much larger than the yield loss attributed to O_3 in either treatment. This uncertainty is unlikely to be resolved until better experimental designs with more precise specifications of the degree of water stress are developed.

O_3 and water-use efficiency

For irrigated crops where water supplies may be limited or expensive, it is important to know whether O_3 would influence the water-use efficiency (WUE) and consequently the amount of water required in a season. The study found that O_3 did not close stomata, but rather affected CO_2 assimilation. A significant increase in leaf construction cost was reported, indicating that both carbon and water are used inefficiently by plants exposed to ozone. In the combination of O_3 and CO_2, there was initially no significant effect of O_3 (i.e., the response of WUE was similar to that observed in elevated CO_2), but as growth progressed, the WUE was significantly reduced in the combination treatment compared to in elevated CO_2 alone. This may be a response to accelerated senescence induced by O_3 exposure.

In total, relatively severe water stress may reduce the yield losses attributable to O_3 in some crop species but, in general, the yield loss resulting from the water stress outweighs the benefits of the O_3 protection. For well-watered crops, O_3 exposure may change the water-use efficiency compared with crops in clean air, but the direction and magnitude of change probably depend on exposure and genetic factors.

Ozone and UV-B radiation

Krupa and Kickert (1989) suggested that interactions might involve episodic exposure to the two stresses, so that peaks of O_3 would coincide with lower UV-B irradiance and vice versa. Runeckles and Krupa (1994) have developed this concept further, arguing that, as tropospheric O_3 increases, UV-B irradiance at the surface is reduced, because the O_3 absorbs some UV. As mentioned earlier, many investigations of plant responses to UV-B are seriously flawed by inadequate specification and/or excessive UV exposure. The ozone treatments resulted in reductions of photosynthesis, accelerated senescence, and reduced yield, in agreement with many other published studies.

Sulphur dioxide on Cotton

High concentrations of sulphur dioxide can produce acute injury in the form of foliar necrosis, even after relatively short duration exposure. However, such effects are far less important in the field than chronic injury, which results from long-term exposure to much lower concentrations of the gas and is essentially cumulative in nature, taking the form of reduced growth and yield and increased senescence, often with no clear visible symptoms or else with some degree of chlorosis. The effects of a given dose of sulphur dioxide can be modified by prevailing environmental conditions. Conversely sulphur dioxide can also modify the response of plants to other environmental stresses, both biotic and abiotic, often exacerbating their adverse impacts.

The sensitivity of cotton to SO_2 was affected by two factors which influenced the rate of growth; namely, light and temperature. When either lower temperatures or smaller photon flux densities were employed to reduce the rate of growth, there appeared to be similar increases in sensitivity to SO_2. It is therefore postulated that growth rate *per se* may be a determining factor in the sensitivity of this species to SO_2. The dry weight of the roots was consistently reduced by SO_2 throughout the range of growth conditions employed here; there was thus no evidence that environmental conditions altered the sensitivity of the roots to SO_2 stress. Effects of SO_2 on shoot: root ratio therefore depends on environmental conditions.

□□□

Chapter 20

Climate Change on Horticultural Crops

Climate plays a major role in crop distribution while weather in crop productivity. Climate decides the crop habitat while crop habit depends on genotype and phenotype. By genotype, it is meant the heredity received by the organism, which is relatively constant throughout life. Phenotype refers to the appearance of the organism - what it looks like - which is subject to change throughout life (Johannsen, 1903). The crop phenology such as flushing, flowering, fruit development and harvest time are sensitive to vagaries in weather and differ from crop to crop. The phenological events undergo several abiotic and biotic stresses or environmental stresses as in cashew flowering during the process of crop growth and development. The effect of environmental stress on phenology could adversely affect the final yield in the same year or the following years depending upon type of crop in plantation crops.

The habitat of plantation crops is different as they are perennials traditionally grown in the humid tropics under rainfed conditions. Within the tropics, crops like coffee, cardamom and tea prefer cool-temperate climate as they grow along the high ranges, where mean annual temperature varies 14 to 20°C while coconut, cashew, cocoa and rubber are seen cultivated in mid- and-high lands, where the annual mean temperature revolves around 25 to 29°C. Interestingly, black pepper is in the buffer zone as an intercrop between warm and cool climates, stretching between tropical and temperate climates within the tropics due to its adaptability.

The response of plants to diurnal/seasonal fluctuation in temperature is known as thermoperiodism. A number of physiological processes, including germination, stem elongation, fruiting, floral development and increase in frost-hardiness proceed most satisfactorily under a rhythm of alternating temperatures. The tomato is sensitive to night temperatures. The optimum temperature may be 26-30°C in young seedlings. As the plant grows older, it prefers lower temperature and the optimum night temperature may range 13-20°C, depending on the variety. The fruit set is abundant with day light temperature of 26.5°C and night temperature of 15-20°C. A high optimum daytime temperature is effective because it increases photosynthetic rate. Tomatoes grow mainly at night and with temperatures above 18°C and translocation of sugar become a limiting factor. Actually, a little fruit set is observed in the field with night temperature above 22°C or below 10°C. In India also, more tomatoes could be produced/plant and fruit size is high if harvested during winter when compared to that of other warm seasons during which temperature may be a limiting factor.

In okra, the fruit yield is directly related to number of heat units. The fruit yield is high if number of heat units is high irrespective of time of sowing and vice-versa. It clearly indicates that okra produces better fruit yield if sowing is taken up from February to April and from September to October.

Plantation Crops

Cashew

Cashew shows a tendency for late flowering and fruiting irrespective of latitude depending upon the altitude. Being a tropical crop, cashew has the ability to thrive in high temperatures. In its natural habitat, it extends into the semiarid regions like Northern Mozambique where daily maximum temperatures exceed 40°C. It is similar in the East Coast of India also, where cashew is grown. It appears that cashew responds to phototropism as it produces better under sunlight. A delay of one week to 10 days in all biotic events of cashew like bud break and flowering may also be noticed towards north direction within a cashew tree. Cashew is not only photosensitive and thermosensitive as the biotic events respond to low minimum and high maximum surface air temperatures. It seems that a temperature range between 15 and 35°C may be the optimum for better growth and production of cashew. In the humid tropical monsoon climates, there is a delay of six days in cashew flowering at every 1° of North latitude and for every 100 m. of altitude, the delay in cashew flowering is three days.

Night temperature between 15 and 20°C, cloudiness and relatively dry weather with the afternoon relative humidity between 40 and 60% may be optimum for triggering pest population of tea mosquito bug in cashew. Tea mosquito bug is a menace over the West Coast and coincides with flushing and flowering time of cashew.

Cardamom

The surface air temperature and its range were high (5.3 - 15.9°C) over Karnataka, followed by Kerala (5.2-11.7°C) and Tamil Nadu (6.9-10.7°C) across the cardamom tract of Western Ghats. Majority of locations showed increasing trend in annual maximum temperature and it was true in South West monsoon and post monsoon periods while declining during winter and summer. Interestingly, an increasing trend was noticed in majority of locations in case of minimum temperature also. The mean annual maximum temperature across cardamom tract is the lowest (24.1°C) over Tamil Nadu and the highest (26.9-30.0°C) over Karnataka while intermediary over Kerala (24.6-27.4°C). The minimum temperature ranges from 15.4 to 17.8°C, dropping down to 13.0°C over Tamil Nadu occasionally.

Coffee

The normal annual rainfall across the coffee growing tract varies widely between 1400 and 4000 mm. About 80% is recorded during monsoon months of June, July, August and September. Number of rainy days varies between 20 and 27 in a month during peak rainy months and July is the rainiest month, followed. The mean monthly maximum temperature is always less than 30°C while the minimum temperature revolves around 24°C and thus the surface atmospheric temperature range is 6-7°C only. The crop is also sensitive to wind disturbances and light conditions. Arabica and Robusta, popularly grown across the West coast, respond differently to various weather elements. The two cultivars are also well demarcated with elevation of growing environment as Arabica prefers high altitudes when compared to Robusta, indicating that Arabica is relatively tolerant to low temperature and humidity when compared to that of Robusta. Growth of coffee across coffee tract appears to be affected when exposed under open conditions due to high temperature during summer and low temperature during winter. It requires a certain degree of shade for its better performance (Table 9).

Table 9. Weather effects on Arabica and Robusta coffee across the Western Ghats

Climatic factors	Arabica	Robusta
Lack of blossom showers	Absence of rains in March-April	Absence of rains in March
Rain on opening day of blossom	Partial to complete failure	Partial failure
Hail storm	Injure the floral and Vegetative phase	Partial failure
Backing showers	Absent in May-poor crop set	Absent in April - poor crop set
Severe western exposures (with lowering of sub-soil-soil moisture)	Partial to complete	Partial to complete
Excessive wetness and water logging	Partial to complete	Partial to complete
Wind	Both are sensitive to wind disturbance	

Tea

In tea, there is a time lag between the physiological function of tea bush and the vegetative harvest. It is about five weeks in the pruned year and four weeks in other years. As such, the weather conditions of previous month are generally correlated to current month's harvest. About 50% reduction in green tea leaves was noticed in April 2004 when compared to 2003 and 2005 in Himachal Pradesh due to increase in maximum temperature of the order of 2.1 to 7.9°C in March (Prasad and Rana, 2006).

Cocoa

Rainfall preferably between 1500 and 2000 mm annually with a dry season of not more than three months with less than 100 mm/month and temperature varying between 30-32°C as mean maximum and 18-21°C mean minimum with an absolute minimum of 10°C are the climatic boundaries desirable for growing cocoa. The loss of apical dominance in cocoa is seen constantly at a temperature of above 32°C. Loss of apical dominance means development of side shoots of plant. A decline of 39% in annual cocoa yield was noted in 2004 when compared to that of 2003 due to rise in maximum temperature of the order of 2-3°C from 14th January to 16th March, 2004 along with prolonged dry spell. Such trend was noticed whenever summer temperature shot up by 2-3°C when compared to that of normal maximum temperature of 33-36.5°C. Adversity of weather aberrations on rhythm of normal growth of cocoa is reflected in yield after a lag period of 4-5 months.

Coconut

Coconut is a tropical tree plant as its distribution in the world is confined almost entirely to tropical zone. In fact, it is highly adaptable to a variety of environmental conditions though it does exhibit some growth preferences. Coconut palms show the following characteristics under severe soil moisture stress depending upon duration and intensity of drought (Rao, 2002).

- Withering and mortality in young seedlings under poor management
- Drooping, wilting and drying of lower whorl of leaves
- Breakage of leaves at petiole or just above it
- Spindle leaf breaking which lead to mortality in senile palms under conditions of poor management
- Abortion of spadices starts from October/November onwards
- Button shedding and immature nut fall
- Nut size decline and
- Finally decline in nut yield in subsequent year up to 50% depending upon type of management and genotype.

The coconut palm likes equable temperature neither very hot nor very cold. The optimum mean annual temperature for better growth and maximum yield is 25-27°C with a diurnal variation of 5-10°C. The coconut palm grows under profuse irrigated conditions where the range of variation in temperature is considerable as seen in Tamil Nadu, Karnataka and Andhra Pradesh. On the West Coast of India, the monthly mean minimum temperature does not fall below 20°C and the mean monthly maximum rarely goes above 33°C. In Malaysia, maximum temperature varies from 30 to 32°C on the coast and 34 to

37°C in the inland. The diurnal variation in temperature is similar to that of 5-10°C. High temperature causes drying up due to desiccation of inflorescence in primordial stage. These would normally open 16 months later in monsoon season and because of their abortion, the leaves, which subtend them, show up barren axils. Abortion of spadices is influenced by high temperature prevailing during hot weather season. The second phase of nut development in coconut is sensitive to heat units since it is the critical phase, which finally decides nut size in coconut. If the number of heat units during second phase of nut development (4-7 months old nut), the final nut size may be diminished. Due to high maximum temperature and low humidity, oil content is relatively less in non traditional areas. It clearly indicates that number of heat units influence crop duration, nut size, oil content and nut yield to a large extent.

Fruit Crops

Mango

Mango is essentially a tropical plant. It can grow at temperatures as low as 0°C and as higher as 45°C, the ideal temperature range for mango is 24-30°C during the growing season along with high humidity. At temperatures below 10°C and above 42°C, growth is affected. Young trees and actively growing shoots are likely to be killed at -1°C. Flowers and small fruits are damaged if temperature drops below 4.5°C for a few hours. High temperatures during the period of fruit development hasten maturity and tend to improve fruit quality. A cool and dry period, generally during winter, that slows or stops vegetative growth is essential in inducing flowering. Precipitation of 890 to 1015 mm in a year, if well distributed is ideal. Rains at flowering are harmful for mango causing total crop failure sometimes. Washout of pollen grains during flowering has been reported by many workers from different parts of the world. During rainy days, pollinating insects remain dull and effective pollination cannot occur. After continued moist weather, severe attack of mango hoppers and certain fungi cause heavy shedding of flowers and fruits and often result in crop failure. Pre-harvest low light intensity due to cloudy weather reduced content of ascorbic acid and sugar in fruits.

Citrus

Citrus plants grow well in tropical and subtropical climate. Very high temperatures are detrimental. The average temperature for sound growth is 16-20°C. Atleast 700 mm of rainfall is necessary which is well distributed. It has been observed that fruits grown in moist climate tend to have thinner peel and more juice than those grown in drier climate. Best quality fruits grow in semiarid, subtropical regions with less than 500 mm of rains, with supplemental irrigation. The fruit acquires a good colour with warm days and cool nights even while growing on trees compared to humid tropics where even ripe fruits remain green. Soil moisture stress coinciding with maturation improves TSS and reduces acidity. Trees attain larger canopy in tropical climate than in subtropical climate. Cool weather in subtropics and moisture stress in tropics are known to condition a major part of the shoots to flower at one time during the year. More pronounced influence of temperature is seen on maturation of fruits.

Guava

It grows best with an annual rainfall around 1000 mm restricted between June and September. However young plants are susceptible to drought and cold. Guava is widely distributed with highest productivity in Gujarat although best quality fruits are produced in Uttar Pradesh (Bose *et al.*, 1999). Among the tropical and subtropical fruits, guava is more tolerant to environmental stresses. The optimum temperature lies between 23°C and 28°C.

Custard apple

During ripening, temperatures above 40°C and below freezing point and low humidity are harmful. No fruit set occurs in northern India due to high temperature and dry condition. For the growth and fruiting of custard apple, an annual rainfall of 800 mm is considered adequate. Although plants can tolerate extreme climatic conditions, yet for good fruiting, high humidity, occasional rains and warm temperatures are required (Bose *et al.*, 1999).

Aonla

Its cultivation is best suited in places with dry summer and frost-free winter. It is adaptable to a wide range of climatic conditions from sea coast to an altitude of 1800m. It has been experienced that in the event of frost, aonla crop is severely affected and in young budded plants, it has been observed that frost affects the scion portion and young grafts are dried. The fruits become whitish in colour, water starts oozing out of them and subsequently dry and turn black (More and Rakeshbhargava, 2010).

Tamarind

Tamarind grows well in tropical climate wherein summers are hot and dry and winters are mild. It is not successfully grown in temperate and arid climate. It is a light demander and cannot be grown in shade. It is drought resistant but susceptible to frost (Singh, 1992).

Vegetable Crops

Tomato

It will not perform well if the temperature goes above 35°C. Similarly its performance will not be good when the temperature goes below 15°C. Between 20 and 25°C, this crop will exhibit its maximum potential.

Brinjal

This crop grows very well in a temperature range of 25°C to 30°C. Very low temperature especially below 15°C will affect the growth of the plant as well as fruit quality. It cannot withstand heavy rains especially during flowering and fruit set.

Bitter gourd

When the temperature ranges from 25°C to 30°C, growth is normal and yields are high. When the temperature is less than 18°C, vine growth is slowed down leading to late flowering and poor yield. When the temperature goes above 36°C, there is a poor production of female flowers and hence the yield is also poor.

Cucumber

Cucumber requires relatively shorter growing season with a warm climate. When the light intensity and duration, and temperature are high, it will favour the production of male flowers and reduce that of female flowers. It is more sensitive to fluctuations in light and temperature than other cucurbits.

Pea

Pea thrives best in cool weather but does not tolerate very hot weather as well as frost. The optimum temperature for germination and initial growth is 18-20°C. But for further growth, flowering and pod set, a temperature range at 12-15°C will be ideal. If the temperature is high, then there will be a drastic reduction in yield.

Cabbage

Most of the cabbage cultivars could be grown under moderate winter conditions. The optimum temperature for growth and heading is between 15°C and 20°C. The growth is arrested when the temperature exceeds 25°C. Seed germination is normal at soil temperature of 12°C to 16°C.

Cauliflower

Since cauliflower has wide varietal variations, it needs specific temperature and photoperiod for better curd development. The optimum temperature for cauliflower ranges from 20° to 25°C in September-October and 5 to 10°C during December-January. At higher temperature, the curds become ricey, leafy, loose and yellow resulting in poor quality.

Based on temperature, the Indian cauliflower is grouped into two. One is the tropical types and the other the annual temperate types called as 'erfurt' or 'snow ball' types. The tropical types form curds at 20°-25°C. The tropical types are resistant to water-logging and heat.

KnolKhol

It is a cool season crop. Ideal temperature for growth and yield is 10°-20°C. Continuous hot weather will hinder the development of knobs.

Carrot

Being a cool season vegetable, it suffers at high temperature. The optimum temperature is 15°C. Germination is good at a temperature range of 7.2° to 23.0°C. It grows well under a wide range of rainfall.

Radish

Radish requires a cool and moderate climate. The ideal temperature is between 10°C and 15°C. During hot weather, the roots become tough and pungent before reaching the edible size.

Beetroot

Beetroot can tolerate both cool and warm climate. It can even withstand some freezing condition. The optimum temperature for better growth is 18°C to 21°C. If the temperature drops down to 10°C for a couple of weeks, bolting occurs and this affects root development.

Turnip

This root crop needs a cool climate and tolerates frost to a certain extent. The optimum temperature is 18-23°C and high temperature will affect the quality of roots.

Potato

The temperature during the growing season is the most important factor influencing the yield. It is propagated vegetatively by whole tuber or a part of it with a bud. For sprouting and initial vegetative growth, a temperature range of 22-24°C is the best. But for tuberization, 18-20°C will be optimum. If the temperature goes above 30°C, then the tuber formation is affected severely. It is sensitive to photoperiodic influence and short days accelerate tuberization. The growth of the top is accelerated by long days and high temperature. So long days and slightly higher temperature will encourage early vegetative growth and in turn help to produce more carbohydrate. As the crop enters into the tuberization period, there should be short days (but more light intensity) and low temperature (but not freezing temperature) viz., cool summer with bright sunlight. This condition will be very ideal for initial better growth of haulms and early tuberization and rapid bulking which in turn results in very high tonnage of tubers.

Yam

The optimum temperature ranges from 30-35°C. When the temperature goes far above especially coupled with dry conditions, growth is adversely affected. Similarly, when it goes down below 20°C, the growth becomes poor. They cannot withstand any degree of frost.

Onion

The optimum temperature is 13°C to 24°C before bulbing and 16-21°C during bulbing. In the young stage, it can even tolerate freezing temperature.

Spinach

It grows very well in a temperature range of 15.5°C - 18.5°C. However, for germination of seeds, a temperature of 10-15°C will be ideal. At higher temperature and long days, the plants tend to flower with the result the quality of the leaves get affected; freezing temperature will cause injury to young plants. At higher atmospheric humidity, the leaves will be more succulent and tender.

Lettuce

The climatic requirements of head lettuce are more critical than that of leaf lettuce or crisp/heading type which is commercially cultivated.Temperature required for germination is 7.2°C to 23°C. But the optimum temperature for growth and formation of head will be 10°C - 18.3°C. When the temperature exceeds 21°C, heading is prevented and bolting occurs. Higher temperature causes bitter flavour and seeding.

❑❑❑

Chapter 21

Climate Change Mitigation Through Crop Breeding Strategies

Introduction

- 1/3 of World Population lives in areas where water is scarce.
- Climate change enhanced the frequency and intensity of drought in sub-tropical areas of Asia and Africa.
- Himalayan glaciers feeding to the Asia largest rivers (Ganges, Indus, Brahmaputra, Yellow river) may disappear by 2035 due to rise in temperature (UN climatic report).

Estimated potential yield losses in crop plants due to abiotic stress

- 17% due to drought
- 20% due to salinity
- 40% due to high temperature
- 15% due to low temperature
- 8% by other factors

Key traits identified for breeding for drought tolerance

- Phenology
- Root shoot association

- Early vigour
- Root density
- Water use efficiency
- Root depth
- Osmoregulation
- Low stomatal conductance
- Sugar accumulation
- Canopy temperature depression

Putative traits for drought tolerance

- Deeper, thicker roots – To explore a greater soil volume
- Root pulling resistance – Root penetration into deeper soil layers
- Greater root penetration ability - To explore a larger soil volume
- Osmotic adjustment – To allow turgor maintenance at low plant water potential
- Membrane stability – Allows leaves to continue functioning at high temperature
- Leaf rolling – Reduce transipiration
- Leaf relative water content – Indicates maintenance of plant water status
- WUE – indicates greater dry weight gain per unit of water lost by transpiration

Breeding approaches

Three important elements of drought characterization for successful breeding of stress tolerance are

- Timing
- Duration
- Intensity of the stress.

Phenotypic performance of a plant/line/population is determined by (1) its genotype, (2) its environment and (3) the genotype x environment interaction. The environment of an individual may be simply defined as the sum total of all the factors other than the individual concerned. The various factors of the environment are called biotic or abiotic depending upon their biological/ nonbiological nature. In an optimal environment, there is no interference by any environmental factor with the complete expression of the genotypic potential of an individual/line; such an environment is, therefore, also called stress-free environment. When some factors of the environment interfere with the complete expression of genotypic potential, it is called stress. Stresses are also classified as biotic (pathogens, pests, weeds, etc.) and abiotic depending on their biological/abiological nature. It should be realized that an optimal environment is an ideal. That is rarely, if ever, achieved, and it is difficult to imagine a stress-free area for commercial cultivation of any crop.

Breeding Methods for Drought Resistance

Breeding methods for drought resistance are the same as for yield and other economic characters. Breeding for drought resistance refers to breeding for yield under soil moisture stress condition. In other words, it refers to yield improvement in environment represented by water deficit. Four breeding methods, viz. 1) Introduction, 2) Selection, 3) Hybridization, 4) Mutations, are commonly used for development of drought resistant crop cultivars. In self pollinated crops, introduction, pure line selection, mass selection, hybridization (Pedigree method and backcrossing) and mutation breeding are used. When drought resistance is available in an exotic variety, such variety can be introduced and after thorough testing, if found suitable, can be released in the new area.

When the drought resistant genotypes are available in the land races or mixed populations, either pure line selection or mass selection is adopted. When the resistant genes are available in the germplasm or wild species, the breeder has to resort to hybridization. Mutation is used when the desired genes are not available in the germplasm. Sufficiently large segregating populations have to be grown to select for drought resistance. In maize, drought tolerance is controlled by no less than genes pairs. Simple screening was effective in selection of drought resistant characters in one genotype, screening of material under drought conditions, use of large population in yield test and testing at several locations will help in selection of superior lines.

In cross pollinated crops, the most commonly used methods are mass selection, backcross, hybridization of inbreds to develop hybrid cultivars, recurrent selection and formation of synthetic cultivars. Mass selection is partially effective because there is no control of pollination. The backcross method is equally successful in self and cross pollinated species. Heterosis is used in both species. However, hybrid vigour is more successful in cross pollinated species than in self pollinated ones.

Breeding for Drought Resistance

There are several environmental factors that have adverse effects on normal growth and development of crop plants. Such factors include deficiency or toxicity of minerals, moisture deficit and low temperature, soil salinity and alkalinity,soil acidity and environmental pollutants. These factors are known as abiotic factors. The stress or adverse condition caused by such factors for growth and development of crop plants are referred to as abiotic stress or environmental stress. Crop plants often suffer from abiotic stresses resulting in significant reduction in both yield and quality.

In some areas, the soil and weather conditions are so much unfavourable that cultivation of certain crops becomes very difficult and non profitable. Under such situation, it is essential either to develop stress resistant cultivars or to modify the environment is the most practical way of solving such problems.

Drought Resistance

Drought refers to the condition of soil moisture deficiency or water scarcity. Soil drought is more common in the arid and semiarid tropics and in the areas of steep slope. Thus, desert areas are more prone to drought conditions. The ability of crop plants to grow, develop and reproduce normally under moisture deficit conditions is referred to as drought resistance. In other words, drought resistance refers to survival of plants under water deficit or scarcity conditions without injury. Improvement in the drought tolerance ability of a plant is known as drought hardening.

Mechanisms of Drought Resistance

There are four different mechanisms which help in survival of plants under moisture deficit conditions. There are: 1) Drought escape,

2) Drought avoidance, 3) Drought tolerance, and 4) Drought resistance. The last one refers to true drought resistance. These are briefly described below:

1. Drought Escape

The simplest way of survival under drought conditions is to escape drought. Generally, drought occurs either in the mid or late-crop season. Drought escape is most common in case of plants grown in desert regions. They complete their life cycles in 4 to 6 weeks. Drought escape also plays an important role in some crop plants. For Example, yields of early varieties of wheat, sorghum, maize, and rice are less affected by severe drought than late maturing ones. All these crops have determinate growth habit. In spring wheat, late maturing varieties give higher yield than early types especially when drought occurs early in the season and is over before anthesis.

2. Drought Avoidance

Drought avoidance refers to ability of the plant to maintain a favourable internal water balance under moisture stress. In other words, plants which avoid drought retain high water contents in their tissues. Drought avoidance can permit a longer growth period in the crop through reduced water use or increased water uptake. However, drought avoidance leads to reduction in photosynthesis and thereby reduction in the growth of aerial parts. It leads to increase in root development and therefore, is more important than drought tolerance. In cereals, drought avoidance operates during vegetative phase, while tolerance operates during reproductive phase. Drought avoidance mechanisms are of two types. First those which reduce water loss through transpiration. Such features include stomatal characteristics and shape, size and orientation of leaves. The second, those which maintain water uptake during drought period.

3. Drought Tolerance

The ability of crop plants to withstand low tissue water content is referred to as drought tolerance. Drought tolerance is more desirable because the crop can produce more yield at lower water potential. In cereals, drought tolerance generally operates during reproductive phase. Tolerant cultivars exhibit better germination, seedling growth and photosynthesis. Drought tolerance differs from drought avoidance in several aspects (Table 10).

4. Drought Resistance

Drought resistance is the sum of drought avoidance and drought tolerance. In other words, drought resistance refers to the ability of crop plants to give good yield under moisture deficit conditions. Drought resistance is measured in terms of various mechanisms associated with drought tolerance and yield under soil moisture deficit.

Table 10. Difference between Drought Avoidance and Drought Tolerance

S.No	Drought Avoidance	Drought Tolerance
1	Plants maintain favourable tissues water content.	Plants do not maintain favourable tissue water content.
2	It reduces photosynthesis and increase root development.	Better seed germination, seedling growth and photosynthesis.
3	In cereals, it operates during vegetative phase.	In cereals, it operates during reproductive phase.
4	Plants can not withstand low tissue water content.	Plants can withstand low tissue water content.
5	It involves various morphological and anatomical features of plant which reduce water loss through transpiration.	It generally involves those characters which support for better photosynthesis under drought conditions.

Basis of Drought Resistance

Drought resistance is associated with various morphological and physiological features of crop plants. Morphological characters which are associated with drought resistance include earliness, shape, size and structure of stomata size, number and orientation of leaves; presence of cuticle; waxiness on leaf lamina and stem, rooting pattern, growth habit etc. Various physiological characters which are related to drought resistance are photosynthetic rate, transpiration rate, osmotic concentration stomatal conductance leaf temperature etc.

Source of Drought Resistance in Plant Breeding

There are three main sources of drought resistance in crop plants: 1) Cultivated varieties, 2) Germplasm collections, and 3) Wild relatives and wild species. Transfer of drought resistance is easy from cultivated variety and germplasm of cultivated species, because such material can be easily used in the breeding programmes.Moreover, there is no problem of cross incompatibility. When the source of drought resistance is a wild species, the transfer of resistance poses several problems such as cross incompatibility, hybrid inviability, hybrid sterility and linkage of several undesirable genes with desirable ones. Wild sources of drought resistance have been reported in wheat, sugarcane, tomato, and several other crops.

Limitation of Drought Resistance Plant Breeding

Breeding for drought resistance has several advantages. It helps in extending cultivation of several field crops in drought prone areas. Drought resistant varieties are more stable to environmental fluctuations and are expected to stabilize crop production under conditions of water stress. Thus, drought resistance breeding is useful in many ways. However, there are some limitations of drought resistance breeding which are given below:

1. Breeding for resistance to various types of stress generally leads to reduction in yield due to strict selection pressure for resistant genes. Breeders are fortunate if any drought resistant line is equal to the best parent in yield. Sometimes, simply inherited characters associated with stress resistance have adverse effect on yield potential.
2. In Spring wheat, a cultivar which gave the highest yield under normal moisture level gave lower yield than many cultivars where grown under drought conditions. Thus, selection for wide adaptability is a selection for medium or even low yield.
3. Genes for drought resistance are mostly found in the germplasm lines which are generally unadapted types. Utilization of unadapted lines poses problems of linkage between desirable and undesirable characters.
4. In some cases, the drought resistant genes are available in wild species. Transfer of resistant genes from wild sources poses several problems such as hybrid inviability, hybrid sterility and linkage between desirable and undesirable genes. This greatly reduces the chances of obtaining desirable combination of favourable genes.
5. Breeding for drought resistance requires cooperation among geneticist, breeder and crop physiologist.

Breeding for Salt Resistance

Salt tolerance refers to the ability of plants to prevent, reduce or overcome the injurious effect of soluble salts present in their root zone. The problem of salinity is of global significance because saline and alkaline soils are found in almost all the countries of the world. In India, more than seven million hectare land is estimated to be salt affected. Higher concentration of salt in the soil adversely affect the growth and development of plants by disturbing various physiological processes.

There are two ways to overcome the problem of salinity. One is to reclaim the salt affected soil and the other is to develop salt tolerant genotypes/cultivars. The first method is very costly, time consuming and short method, on the other hand, the second method is long lasting, more effective and less costly. Genetic differences exist among cultivars for their salt tolerance capacity. Moreover, plants adapt themselves to adverse environments for their survival.

Breeding Approaches

In India, breeding work for salt resistance is carried out at Central Soil Salinity Research Institute (CSSRI), karnal and its various regional centres. In India, two types of approaches are followed for salt tolerance breeding. The two approaches are: 1) Improving yield level of salt tolerant cultivars, and 2) Transfer of salt tolerant genes to high yielding cultivars. In the first, traditional cultivars of salt affected areas are improved for their productivity without affecting their salt tolerance ability. In the second approach, salt tolerance genes from locally adapted (Salt tolerant) cultivars are transferred to high yielding cultivars through hybridization and selection.

The salt tolerance capacity differs from species to species. For example, crops like barley, cotton, sugarcane, oilseeds and grasses are known to be more salt tolerant than other crops. The salt tolerance capacity also differs within a crop based on ploidy level. For instance, in wheat, hexaploid bread wheat is more tolerant to salt than durum (tetraploid) and (diploid) wheat. In Brassica, tetraploid species are more tolerant than diploids. In rice, late maturing, tall and coarse grain varieties exhibit maximum tolerance to salinity conditions. In sugarcane, vast differences for salt tolerance are observed among strains. In Hawaii, salt shock treatment is used for improving salinity tolerance. In sugarcane, moreover, land races are more tolerant to salinity than high yielding cultivars.

Breeders require close cooperation of geneticist, physiologist, biochemist, and soil scientist in developing salinity resistant varieties. The new approaches such as tissue culture and genetic engineering may be rewarding in developing more salt tolerant genotypes.

Molecular Breeding for Drought and Saline Tolerance

Recent research break throughs in biotechnology have revived interest in targeted drought tolerance breeding and use of new genomic tools to enhance crop water productivity. Marker-assisted breeding is revolutionizing the improvement of temperate field crops and will have similar impacts in breeding of tropical crops. Other molecular breeding tools include development of genetically modified crops that can tolerate plant stress. As a complement to the recent rapid progress in genomics, a better understanding of physiological mechanisms of drought response will also contribute to the progress of genetic enhancement of crop drought tolerance.

It is now well accepted that the complexity of the drought syndrome can only be tackled with a holistic approach that integrates physiological dissection of crop drought avoidance and tolerance traits using molecular genetic tools such as MAS, microarrays and transgenic crops, with agronomic practices that lead to better conservation and utilization of soil moisture, and better matching of crop genotypes with the environment.

Achievements in Resistance Breeding

Drought resistant varieties have been developed in several crops in many countries throughout the world. In wheat, drought resistant varieties have been developed in China and USSR. All these varieties have been developed through induced mutations. In India, drought resistant varieties have been evolved in cotton. rice, sugarcane and several other crops. In cotton, drought resistant varieties have been developed through induced mutations. In India, breeding work for salinity resistance is mainly carried out at Central Soil salinity Research Institute, Karnal (Haryana). Several salt resistant varieties of rice, barley, wheat and sugarcane have been released

A drought tolerant variety of rice, called Sahbhagidhan, was released for commercial cultivation in 2009. This variety yields upto 35 q/ha under drought situation. Another rice variety, Swarna Sub-1, has been developed to tolerate submergence under flash flood situations. This variety can tolerate upto 10 days complete submergence condition and produce 65 q/ha. It has been developed by transferring, using MAS (marker-assisted selection), into variety Swama (MTU-7029) the gene *subl* for submergence tolerance from a land race collected from Orissa.

❑❑❑

Chapter 22

Mitigation of Climate Change Through Crop Breeding Strategies for Grain Yield and Quality

Introduction

Changes in climate can be expected to have significant impacts on crop yields through changes in temperature and water availability.The yield and quality of food crops is central to the well being of humans and is directly affected by climate and weather.Initial studies of climate change on crops focussed on effects of increased carbon dioxide (CO_2) level and/or global mean temperature and/or rainfall and nutrition on crop production.However, crops can respond nonlinearly to changes in their growing conditions, exhibit threshold responses and are subject to combinations of stress factors that affect their growth, development and yield. Thus, climate variability and changes in the frequency of extreme events are important for yield.

Crop yields are affected by many factors associated with climate change, including

1. Temperature
2. Rainfall
3. CO_2 concentration in the atmosphere
4. Extreme weather events
5. Climate variability.

Climate Change and Response Strategies

Mitigation

Research for mitigating climate change is aimed at countering or reducing greenhouse gas emissions and involves three key approaches:

- Emit less greenhouse gas from all parts of the food chain.
- Lock up CO_2 - Carbon (C) sequestration in soil and vegetation.
- Replace fossil fuel with renewable bioenergy crops.

Plant Response to Abiotic Stress

Primary stresses, such as drought, salinity, cold, heat and chemical pollution, are often inter connected and cause cellular damage and secondary stresses, such as osmotic and oxidative stress. The initial stress signals (e.g. osmotic and ionic effects or changes in temperature or membrane fluidity) trigger the downstream signalling process and transcription controls, which activate stress-responsive mechanisms to re-establish homeostasis and to protect and repair damaged proteins and membranes. Inadequate responses at one or more steps in the signalling and gene activation process might ultimately result in irreversible changes in cellular homeostasis and in the destruction of functional and structural proteins and membranes, leading to cell death.

High Temperature (HT) tolerance

Plants are adapted to limited range of temperatures. If the temperature drops below 15^{0}C plants experience low temperature stress and if it is above 45^{0}C, plants are subjected to high temperature stress. Increase of 15-20^{0}C above normal temperature causes deeper modification of growth without being necessarily lethal *viz.*,

- Protein denaturation,
- Enzyme inactivation,
- Reduction in chloroplast's photosynthetic activity

Role of Heat Shock Proteins (HSPs) in higher temperature

Plants have the capacity to interact with the environment in many different ways and to survive under extreme abiotic and perhaps also biotic stress conditions. The response to heat stress (hs) is highly conserved in organisms but owing to the sessile life style it is of almost importance to plants. The hs-response is characterised by (i) a transient alteration of gene expression (ii) by the acquisition of a higher level of stress tolerance (acclimation). The induction of HSP-expression is not restricted to high temperature stress; HSPs are also linked to a number of other abiotic stresses including cold, freezing, drought, dehydration, heavy metal, and oxidative stresses. HSP are molecular chaperones, which either prevent complete denaturation (small HSP: sHSP) or are supporting proper folding (other HSP) of enzymes under or after protein denaturing conditions. Manipulation of the heat stock protein response has the potential to improve common stress tolerance that may lead to a more efficient exploitation of the inherent genetic potential of agriculturally important plants. Another approach is expression of gene involved sugar alcohol like aspartate decarboxylase and glycine betaine that will be used for resistance against heat tolerant.

Role of Aquaporins during stress

Dehydration and salt stress requires change in water flow to allow cells and tissue to adapt to the stress situation. The rate of water flux into or out of the cells is determined by the water potential gradient that acts as the driving force for transport and by the water permeability of the membrane. Aquaporin proteins facilitate osmosis by forming water specific pores as an alternative to water diffusion through the lipid bilayer, thus increasing the water permeability of the membrane. Aquaporins are membranes of a large super-family of membrane spanning proteins, the major intrinsic proteins. In plants, aquaporins localized in the tonoplast are called tonoplast intrinsic protein; while those in the plasma membrane are PIPs.Across different plant species,aqauaporins have very conserved structures and hence are encoded by similar DNA sequence. These functions were inferred from the expression patterns of specific aquaporin isoforms or from the blocking effect of mercury on water transport through plant tissues. There are several reports that aqauaporin genes are induced by dehydration and salt stress; this may trigger greater osmotic water permeability and facilitate water flux.

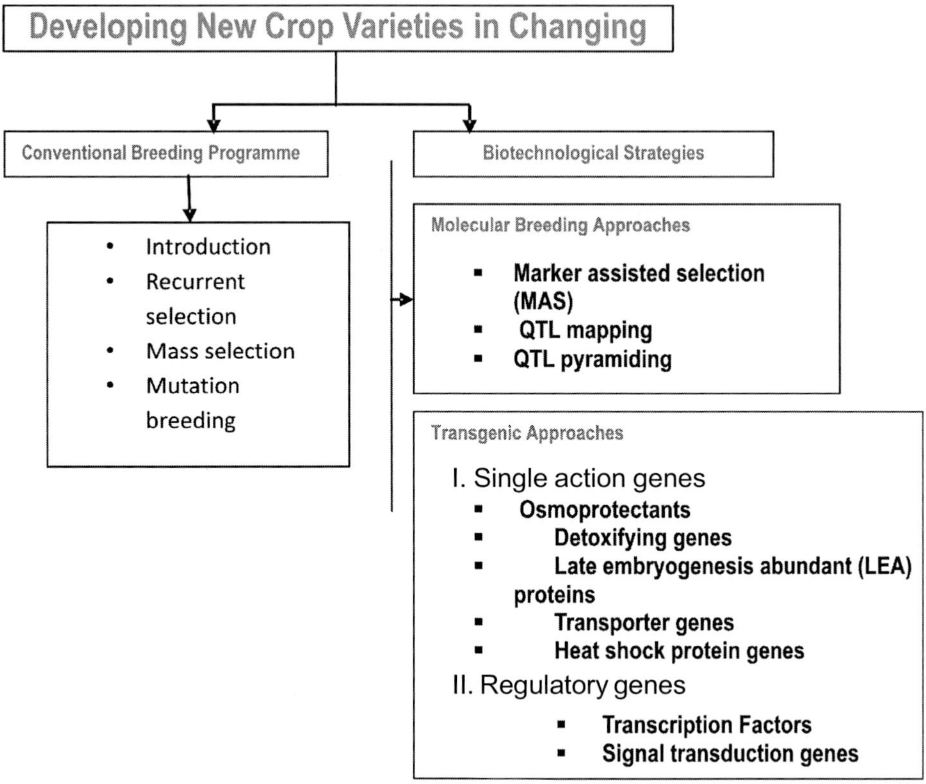

Breeding approaches for salt tolerance

Sustainable food production puts a high demand on breeding more salt tolerant crops. Breeding of crops tolerant to soil salinity and yield well in salt affected soils, and employing crop management practices to counter salinity, have been proposed to maintain crop productivity. The generalized step for conventional breeding approaches for salt tolerant are:

1. Screening germplasm collection for donors of salt tolerance
2. Crossing donors with an elite line and advancing the F_1hybrid to all F7/8 generations while selecting the elite characters.
3. Selecting for tolerance starting about F_4 generation.

Breeding approaches for heat tolerance

In the development of high temperature tolerant genotypes of crop plants, it is necessary to improve plant productivity under high temperature stress environment. The quantification of high temperature tolerance and characterization of its genetic control are necessary for germplasm enhancement efforts. Breeding against heat injuries in crop plants, tolerant breeding materials, such as tolerant cultivars, inbred, advanced lines, and wild relatives are most essential.

Heat stress resistance

Heat stress resistance may be defined as the ability of some genotype to perform better than others when they are subjected to the same level of heat stress. The various machanisms of heat resistances may be grouped into two categories. 1. Heat avoidance and heat tolerance (Table 11).

Table 11. Different mechanisms of heat stress resistance in plants

Mechanisms	Contributory processes	Consequences Remaks
Heat avoidance	Transpiration	Cooling
	Leaf reflectance due to	
	1. Pubescences	Reduced light interception by leaves
	2. Glaucousness	
	Insulation by Bark	Reduce heating
Heat tolerance	Membrane stability Stability of photosystem II Photosynthate translocation Stem- reserve mobilization Osmoregulation	Association with heat hardening and possible Hspsynthesis

Cold resistance

In many environments, crop productivity is limited by low temperatures. When temperatures remain above freezing, *i.e.*> 0^0 C, it is called chilling, while freezing describes temperature below freezing, , *i.e.* < 0^0 C. Chilling and freezing stress and the mechanisms of resistance to them are considerably different. Chilling sensitive plants are typically tropical plants. Chilling stress may be described either at the plant/ organ level orat the subcellular level. But all the effects of chilling stress results from its primary effect on cellular membranes.

Selection of environment

Selection for chilling tolerance can be effective only if appropriate selection pressure is applied. Therefore, control over the stress environment is critical for the success of such programmes. Selection may be carried out 1). in the field in or 2). in a controlled environment, *e.g.*, growth chamber.

Impact of Climate Change on Quality of Cereals and Oilseed Crops

Elevated temperature and grain quality of wheat

Breeding for superior grain quality has become as important as breeding for yield. Increase in temperature can have significant effect on quality of various crops. Over 80% of the nitrogen in the crop is taken up before flowering and stored in stems and leaves. It is deposited in the grain in the form of protein at a rate that reaches its maximum during early part of grain filling. Carbohydrate in the form of starch is also deposited in the grain but its rate reaches a maximum later in the grain filling process. The eventual protein concentration of the grain is thus, a result of the amount of dilution of the protein by the carbohydrate supply. Under heat and drought stress, the natural reaction for wheat will be to store less carbohydrate and conversely store more protein. This happens when the plants are stressed after heading and during grain filling. Increases in daily average temperatures above 30^0 C, for a period of 5 days resulted in weakening of dough properties like dough strength and leaf score accompanied by increase in proportion of gliadin proteins.

In general, grain nitrogen concentration increases with increase in temperature and reduced rainfall. In Australian variety trials extending over 27 years, grain nitrogen concentration was positively associated with the number of above 35^0C during grain filling. Increasing temperature can reduce carbohydrate accumulation more than nitrogen accumulation as the rate of senescence is increased thereby reducing photosynthesis and grain growth. The rate of protein synthesis in the grain is also promoted by warmth more than the rate of starch synthesis such that protein concentration increases. It was observed that high temperatures and low relative humidities were associated with reduced gluten and increased gliadin contents, respectively. During grain filling, these factors, despite increasing grain nitrogen concentration, reduce dough strength, gluten quality and loaf rating.

Elevated CO_2 and grain quality of wheat

There is scarcity of information on the effects of rising atmospheric CO_2 on nutritional value and grain quality parameters, apart from protein concentration. Elevated CO_2 reduces the protein content of grain and flour by 9-13%. Plants grown under elevated CO_2 in open field, open top chamber and FACE resulted in decrease in protein concentration. Protein concentration is positively correlated with bread making due to its impact on dough strength and is a major determinant of grain prices. The total gluten concentration as well as concentration of dry and wet gluten were also decreased under high CO_2.

Elevated CO_2 results in decrease in the concentration of all microelements by 3.7-18.3 % whereas macro-elements are not affected. A proper balance between protein and non protein components of grain determines the quality of wheat flour. Grain grown in high CO_2 produce poor dough of lower extendibility and decreased loaf volume but the physiochemical properties of wheat starch during grain fill are not significantly modified.

Elevated temperature and grain quality of rice

Rice is the most important food crop in the world providing over 21% of the calorific needs of the world's population. The quality of rice is strongly influenced by high temperature prevailing during grain development phase of rice and the impact varies with cultivar and its adaptability change. Warm weather condition (mean temperature >28^0 C) with high solar radiation (> 10,000 cal) without excessive rains during grain filling stage produced the best rice yield with high quality. Aromatic rice requires relatively cooler temperature (25^0 C day and 21^0C night) during crop maturity for better retention of aroma. Amylose content of grain decreased linearly with increase in mean temperature. Elevated CO_2 causes an increase in amylase content.

Elevated CO_2 and grain quality of rice

Marginal increase in the canopy temperature due to increase in CO_2, has been associated with a decrease in yield. Elevated CO_2 could increase potential yield by increasing leaf area and the number of spikelets. However, any yield advantage gained from elevated CO_2 could be negated by high night time temperature. Increased grain yield observed under elevated CO_2 was due to both increased tiller number/ plant and due to bolder and denser grains.

Elevated temperature and quality of oilseed

Oilseed crops like mustard, groundnut, sunflower, castor, sesamum etc., are grown world over under rainfed conditions. These conditions make oilseed crops prone to climate changes. Few studies have been conducted to study the impact of climate change on quality of these crops. Higher rainfall and cooler temperature produced higher yield and oil content.

The effects of high temperature on protein concentration are more than oil concentrations. Oil concentration has shown an inverse relation with protein concentration with respect to temperature change. The fatty acid composition is also known to vary with high temperature with an increase in percentage of monounsaturated fatty acid and decrease in polyunsaturated fatty acid in mustard.

Elevated CO_2 and quality of oilseed

Elevated CO_2 has been known to ameliorate the reduction in carbohydrate content observed under water stress condition in mustard. Also, there was an increase in grain mass and carbohydrate concentration and reduction in crude fibre content. Seed oil content increased and unsaturated fatty acids decreased under elevated CO_2. In sunflower, due to elevated $CO_{2,}$ oil percent increased, but protein percent decreased. Analysis of seed quality traits revealed that 100 seed weight and seed vigour improved in seed harvested from plants exposed to elevated CO_2.

❑❑❑

Chapter 23

Biotechnological Approaches to Improve Crop Adaptation to Climate Change

Developing crops that are better adapted to abiotic stresses is important for food production in many parts of the world today. Anticipated changes in climate and its variability, particularly extreme temperatures and changes in rainfall, are expected to make crop improvement even more crucial for food production.

Importance of Crop Improvement

The intensification of agriculture in many parts of the world over the past five decades, supported by appropriate research institutions and policies, has led to an increase in global food grain production from approximately 850 million tons in 1960 to 2350 million tons in 2007. Although global food production has kept pace with population growth over the past 40 years, almost a billion people, most of who live in the developing world, remain undernourished. It has been projected that global food production must increase by 70% by 2050 to meet the demand caused by this growing global population, increasing incomes and consumption. Food insecurity has also increased in recent times in several regions of the world owing to competing claims for land, water, labour, energy and capital, which is leading to more pressure to improve production per unit of land. Global climate change is likely to increase the problems of food insecurity, hunger and malnutrition for millions of people, particularly in south Asia, sub-Saharan Africa and small islands and may also further aggravate the current trends in land degradation, especially in semi-arid tropical regions. Whereas global temperatures are predicted to increase by 2.5–4.3° by the end of the century, with significant effects on food production and malnutrition, it is also evident that agriculture is currently affected by increasing climate variability, especially temperature. For example, it has been estimated that rising global temperatures between 1981 and 2002 reduced the yields of major cereals by $5 billion per year. Heat waves and drought in Europe in 2003 significantly reduced productivity [e.g. maize (*Zea mays*) yield in northern Italy by 36%]. In 2009/2010, heat waves also affected wheat (*Triticum aestivum*) production in central Asia and extreme flooding affected agricultural production in south Asia. In addition to the challenge of temperature extremes (hot and cold) and drought or water stress as well as flooding associated with climatic variability, the incidence and severity of biotic stresses such as pests, diseases and the invasion of alien weed species are also likely to be greater. Hence crop improvement through biotechnological interventions has gained importance.

Biotechnological interventions

The objective of plant breeding for stress environments is to accumulate favourable alleles that contribute to stress tolerance in a plant genome. Genes that confer stress resistance can be sourced from germplasm collections, including wild relatives of crops that are held in gene banks or organisms that currently live in the habitats of water deficit or excess, extreme temperature and salinity that have evolved to cope with those conditions. Some progress has been made through conventional breeding.

Conventional Breeding for abiotic stress tolerance is constrained:

i) By the complex nature of abiotic stress tolerance (timing, duration, intensity, frequency) and thereby its quantification and repeatability;

ii) Because undesirable genes are also transferred along with desirable traits;

iii) Because reproductive barriers limit the transfer of favourable alleles from diverse genetic resources.

Biotechnology is a viable option for developing genotypes that can perform better under harsh environmental conditions, particularly for (ii) and (iii) above.

For instance, advances in genomics coupled with bioinformatics and stress biology can provide useful genes or alleles for conferring stress tolerance. Superior genes or alleles that have been identified in the same species can be transferred into elite genotypes through molecular breeding. Moreover, by using an approach such as genetic engineering, there is no barrier to transferring useful genes or alleles across different species from the animal or plant kingdoms. As a result, biotechnological approaches offer novel strategies for producing suitable crop genotypes that are able to resist drought, high temperature, submergence and salinity stresses. The biotechnological approaches could be used to develop crops with potentially improved productivity in an environment with high temperature, high CO_2 and high ozone. These included manipulating leaf photosynthesis, photosynthetic partitioning, total biomass production and nitrogen use efficiency (NUE). Improved NUE in crops should lead to reduced fertilizer application and thereby lower emissions of greenhouse gases into the atmosphere. More than 50% of all US greenhouse gas emissions from agriculture are associated with fertilizer application and other cropping practices. Rather than focusing on individual stresses, the biotechnological community should use biotechnological approaches to tackle multiple stresses directly under field conditions.

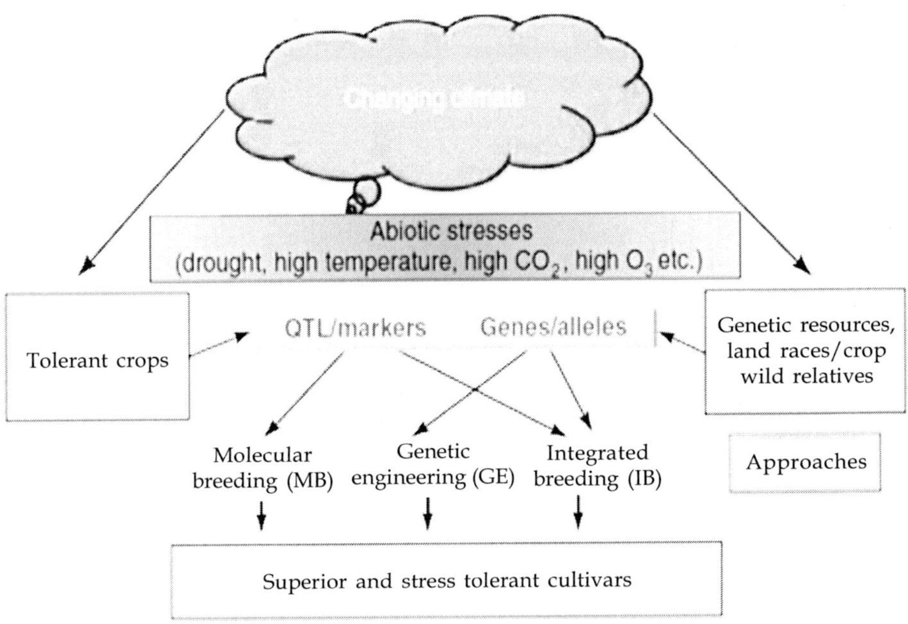

Fig 3. Key biotechnological strategies for improving abiotic stress tolerance

Key Agricultural Biotechnology Approaches: Prospects and Progress (Fig. 3)

Molecular Breeding approach

- The development of genomic resources such as molecular markers including simple sequence repeats, single nucleotide polymorphisms and marker genotyping platforms.
- The development of biparental mapping populations by using genetically and phenotypically diverse parental lines or the selection of a natural population representing diversity for abiotic stress tolerance traits.
- The use of linkage mapping or association mapping approaches to identify the QTLs or markers associated with abiotic stress tolerance-related parameters, such as leaf water retention, high rates of leaf photosynthesis, stomatal conductance, osmotic adjustment and faster canopy and root development.
- The validation of the QTLs or markers in a breeding germplasm that have a different genetic background.
- The use of an appropriate MB approach such as MABC, MARS or GWS to develop superior crop genotypes.

Genetic Engineering approach

- The identification of genes encoding signaling proteins, TFs and effector proteins, and novel stress responsive promoters controlling multiple stress tolerance.
- The identification of genes regulating stomatal opening and closure and stress-induced expression to enhance water use efficiency in crops.
- The genetic transformation and development of elite crop genotypes with tolerance to high temperature stress and other environmental stresses.
- The assessment of promising transgenic lines for multiple stress tolerance under field conditions.
- The deregulation of transgenic lines to enable the release of a superior line or variety.

Biotechnological Tools to transfer genes into plants

- Agrobacterium mediated
- Gene gun/ Biolistic
- Electroporation
- Microinjection
- Chemical mediated
- Lipofection
- Macroinjection
- Fibre mediated DNA delivery
- Laser induced DNA Delivery
- Pollen transformation
- DNA Delivery via growing pollen tube
- Direct DNA uptake by Mature zygotic embryos

Agrobacterium mediated

- Gene of interest isolated from desired source
- Integrate it into agrobacterium
- Coculture plant cells with agrobacterium
- Wounded plant cells secrete special phenolic compounds
- Sensed by Vir genes of agrobacterium
- Leads to activation of all genes of T-DNA
- T-DNA excised from Agrobacterium and ultimately integrate into plant genome.

Gene Gun

- Coat gold or tungsten particles with Plasmid DNA
- DNA is driven by helium blast (old days gunpowder driven pistons)
- Can be used against cells or whole plants (mono or dicots)

Electroporation of protoplasts

Introduction of DNA into the cells by exposing them for very brief periods of high voltage electrical potential, induce transient pores in the plasma membrane. Plant protoplast and recombinant plasmid DNA are suspended in a suitable ionic solution

Chemical mediated transformation

- Chemicals like PEG, Polyvinyl alcohol enhance DNA uptake by protoplast
- Protoplast along with recombinant plasmid suspended in transformation medium along with Mg^{++} and 20% PEG
- 5 min heat shock of 45^0C and immediate cool on ice

Microinjection of protoplasts

DNA solution is injected directly inside the protoplast using capillary glass micropipette with the help of a microinjection assembly.

Validation of gene transfer

- Molecular Marker analysis
- Reporter gene markers
- PCR
- Southern blot hybridization
- RT-PCR
- Northern
- Western

Molecular breeding

Significant advances have been made in the area of genomics over the past ten years. Genome sequences are available now for many crop species such as rice, sorghum, maize and soybean. Further more, the advent of so-called 'next-generation sequencing (NGS)' technologies have made it possible to sequence the transcriptomes or genomes of any species (and for any number of individuals) relatively, quickly and cheaply. As a result, genome sequences have started to become available for less studied crops such as cucumber, pigeon pea and large and complex genome species such as wheat and barley. These genome or transcriptome sequences coupled with genetic approaches can be used for identifying suitable genes conferring stress tolerance that can be deployed in crop improvement either by using MB or GE approaches.

The use of genome sequences to identify genes associated with drought tolerance can be demonstrated by taking the example in sorghum, a species well adapted to drought prone regions. Sorghum genome analysis has indicated that the characteristic adaptation of sorghum to drought might be partly related to the expansion of one miRNA and several gene families. Rice miRNA 169g, up regulated during drought stress, has five sorghum homologs (sbi-MIR169c, sbi-MIR169d, sbi-MIR169.p2, sbi-MIR169.p6 and sbi-MIR169.p7). The computationally predicted target of the sbi-MIR169 subfamily comprises members of the plant nuclear factor Y (NF-Y) B transcription factor family, linked to improved performance under drought by Arabidopsis (*Arabidopsis thaliana*) and maize. Cytochrome P450 domain-containing genes, often involved in scavenging toxins such as those accumulated in response to stress, were found to be abundant in sorghum (326 cf 228 in rice). Expansins, enzymes that break hydrogen bonds and are responsible for a variety of growth responses that could be linked to the drought tolerance of sorghum, occurred in 82 copies in sorghum 58 in rice and 40 each in Arabidopsis and poplar.

The MB approach involves first identifying quantitative trait loci (QTLs) for traits of interest, such as tolerance to abiotic stresses. Until recently, QTLs were identified by linkage mapping, but now association genetics has started to supplement these efforts in several crops. Nested association mapping, which combines the advantages of linkage analysis and association mapping in a single unified mapping population, is also being used for the genome-wide dissection of complex traits in maize.

Association mapping, compared with linkage mapping, is a high-resolution and relatively less expensive approach. In the near future, it is likely to be routinely used for identifying traits associated with abiotic stresses, particularly given the availability of high-throughput marker genotyping platforms. An example of the systematic use of association mapping for drought tolerance is the collaborative project between Cornell University and CIMMYT. After identifying the markers associated with QTLs or genes for traits of interest, the candidate QTLs or genes can be introgressed into elite lines through marker-assisted backcrossing (MABC).

Although MABC has been successful in developing superior genotypes for traits controlled by major effect gene(s) or QTLs, for example bacterial blight and blast resistance in rice, few examples are available for complex traits such as tolerance to drought and heat, which are the key traits that need to be targeted for developing crops that are adapted to low rainfall and high temperature conditions. However, MB has been successfully used in rice, with one major effect QTL each for submergence tolerance and drought tolerance identified and used in this approach. One of the difficulties of developing superior genotypes for abiotic stresses such as drought or heat is that these traits are generally controlled by small effect QTLs or several epistatic QTLs. MABC does not seem to been effective approach for introgressing QTLs such as these, especially because of the large sizes of the backcross populations required to pyramid several QTLs in the same genetic background.

However, two newer MB approaches–marker-assisted recurrent selection (MARS) and genome wide selection (GWS) or genomic selection (GS) - can be used to overcome this problem. The estimated genetic gain that is feasible using MARS or GWS is greater than can be obtained using MABC for transferring or pyramiding superior QTLs or gene alleles for complex traits such as drought or heat tolerance in one genetic background. Although the MARS approach is used routinely in private sector breeding programs, there are no published reports on the use of MARS in public breeding programs. Another comprehensive approach for improving complex traits is based on GWS. Although MABC and MARS require QTL information for complex traits, information on marker–trait associations is not necessarily required for GWS. Basically, GWS deals with the prediction of the genomic-estimated breeding values (GEBVs) of progeny. In this context, there is first a need to have the phenotyping data as well as genome-wide marker profiling on a 'training population'; subsequently, GEBVs can be calculated based on phenotyping and marker datasets. These GEBVs are then used to select the superior progeny lines for advancement in the breeding cycle. Several computational tools are available or are being developed to calculate GEBVs, such as the Best Linear Unbiased Prediction method and the geostatistical mixed model. However, at present there is little information available on the use of GWS in crop plants in public sector breeding programs, although some groups have started to explore this approach in crops such as maize

Genetic engineering

The upsurge of genomic information and the use of associated computational biology tools over the past decade have led to the identification of signalling pathways and regulatory genes and networks controlling complex traits related to environmental stresses. Crop genetic engineering with signalling components and transcription factors (TFs) lead to expression of their target transcriptome that consists of several genes involved in stress adaptation. For example, the enhanced production of the signaling hormone abscisic acid (ABA) by the over expression of the LOS5/ ABA3 gene encoding a Molybdenum Cofactor Sulfurase, required for ABA synthesis, conferred enhanced drought tolerance in transgenic rice plants under field conditions. Similarly, the over expression of the rice AP37 (an APETALA2-typeTF) gene resulted in the enhanced expression of several target genes and produced 16–57% higher grain yield under field drought stress conditions. Hence, transcriptome engineering seems to be promising for the development of abiotic stress-tolerant crops. However, inducible expression rather than the constitutive over expression of TFs is preferable owing to the severe growth retardation and reduction in seed production that can occur even under normal environmental conditions in transgenic crops with constitutive expression of TFs. Nonetheless, several transgenic crops have been engineered using C-repeat binding factors (CBFs) and other TFs without a yield penalty. Transgenic rice plants over expressing Arabidopsis CBF3/DREB1A orABF3 TF showed improved tolerance to drought and salinity without growth retardation. However, only a few crops such as rice, maize and canola, expressing the desired TF and other genes, have been tested under real field stress conditions. RNA chaperones known for their active role, particularly in mediating transcription and translation both in bacteria and plants, have also been shown to increase yield under multiple stresses. For instance, Monsanto researchers showed that bacterial cold shock proteins (Csps) can confer improved stress adaptation in multiple plant species. For instance, CspB codes for and is responsible for an RNA chaperone, which is a commonly occurring protein molecule that binds to RNAs and facilitates their function. The gene was first identified in bacteria subjected to cold stress conditions, and further research has demonstrated that CspB helps plants cope up with drought stress. In maize and rice, CspB works by helping the plant maintain growth and development during times of inadequate water supply.

Recently, a gene encoding aquaporin (NtAQP1) was identified in tobacco and shown to provide protection against salinity stress in transgenic tomatoes. NtAQP1 plays akey role in preventing root/shoot hydraulic failure, enhancing water use efficiency and thereby improving salt tolerance. It simultaneously increased both water use and photosynthetic efficiency in plants. Moreover, the NtAQP1gene, which increases stomatal conductance, might also lower canopy temperature and thereby reduce the level of heat stress experienced by plants. By contrast, decreased stomatal conductance and thereby transpiration by the suppression of farnesyl transferase genes (FTA or FTB) by RNAi in transgenic canola resulted in significantly higher yields compared with controls in a 3-year field trial. To make up for the water loss owing to higher stomatal conductance in the NtAQP1 transgenic plants, pyramiding genes for osmolyte biosynthesis expressed specifically in roots could lead to the growth of deeper roots, potentially enabling water uptake from deeper soil layers. A combination of genes is required to offset the adverse impact of climate variability on plants. Apart from genes encoding effector proteins, signaling proteins and/or TFs, there is a need for promoters to drive transgene expression in a precise and predetermined fashion in specific tissues or plant organs. Appropriate promoters need to be selected, depending upon the gene used, to obtain desirable transgenic plants with high yield stability under stress conditions. For example, as discussed earlier, the constitutive expression of ZmNF-YB2 in maize conferred enhanced drought tolerance. By contrast, transgenic rice plants over expressing OsNAC10 under the root specific promoter RCc3, but not under the control of the constitutive GOS2 promoter, conferred a yield advantage under drought stress conditions in the field. The use of appropriate promoters might enable gene pyramiding through genetic engineering to tackle the issue of tolerance to multiple stresses at different stages of plant growth and development. TFs implicated in more than one type of stress might also be identified. The role of TFs in enhancing agricultural productivity via increased leaf photosynthesis, modified plant architecture and faster plant. Other TFs, such asDREB2A with diverse roles in both biotic and abiotic stresses, could possibly be deployed for engineering multiple stress tolerance along with enhanced yields. However, the key concern about TF transgenics is whether they will perform consistently under drought and/or heat stress conditions in the field.

Marker-assisted breeding (MAB) for drought tolerance

Like tolerance to other abiotic stresses, drought stress is also controlled by many minor genes (polygenes) that have additive effects in their expression. Thus, the loci on chromosomes housing such types of genes are now referred to as quantitative trait loci (QTL). Natural genetic variation of a crop can be exploited either via direct selection under stressful conditions whether simulated or natural or via mapping of QTL (polygenes) and subsequent marker-assisted selection (Table 12).

In cotton, using F_3 families derived from the cross, *Gossypium barbadense* cv. F-177 and *Gossypium hirsutum* were identified a subset of 33 QTLs under water limited regime, i.e., 11 QTL for plant productivity, 5 for some key physiological traits, and 17 for fiber quality.

Table 12. Drought tolerance in different crop lines by molecular breeding approaches

Crop	QTL used	QTL donor line or Cultivar	Recipient line or cultivar	Traits improved
Rice	QTL 9 on Chr. 9	Azucena	Kalinga	Root length and thickness
Pearl millet	QTL on linkage group 2 (LG 2)	PRLT 2/ 89-33	H-77/ 833-2	Introgressed lines showed grain yield and out-yielded test cross hybrids
Maize	Five QTLs located on chromo-somes 1, 2, 3, 8, and 10	Ac7643	CML 247	Under severe drought stress conditions, MAS derived varieties were about 50% better in yield.
Barley	81 QTLs were used,	*Hordeumsp-ontaneum*	*Hordeum-vulgare*	Improved grain yield, and reduced

	out of which 6 were for grain yield			negative impact of drought on grain filling
Sorghum	Stg1, Stg2, Stg3, or Stg4	BT×642	BT×642 RT×7000	Delayed leaf senescence and better grain yield at maturity under drought stress conditions

Engineering crops for enhanced drought tolerance - Transgenic Approach

Transgenic approach is being pursued actively throughout the world to improve traits including tolerance to biotic and abiotic stresses in a number of crops. It is now well established that compatible organic solutes play a central role in plant drought tolerance. However, over production of compatible organic osmotica is one of the prominent responses of plants exposed to osmotic stress and the genes encoding the synthesis of such organic solutes can be engineered to overproduce these solutes in transgenic plants (Table 13). For example, among the many organic osmolytes known to play a substantial role in stress tolerance, glycine betaine (GB), a quaternary ammonium compound, synthesised in response to dehydration stress. The transformed maize plants contained higher levels of glycine betaine and showed higher tolerance to drought as compared to wild-type plants when tested at the initial growth stages.

Table 13. Examples of improved drought tolerance through genetic engineering

Gene	Host plant	Source organism	Traits improved
Beta vulgaris choline monooxy-genase (BvCMO)	Tobacco (*Nicotiana tabacum* L.)	Beet (*Beta vulgaris* L.)	Higher accumulation of glycine betaine in leaves roots and seeds. Transgenic line exhibited improved tolerance to toxic level of choline and drought stress
Choline dehydrog-enase (betA)	*Zea mays* L. inbred line DH4866	*E. coli*	Higher accumulation of glycinebetaine; Tolerant to drought stress at germination
Triticum aestivum salt tolerance gene (TaSTRG)	Rice (*Oryza sativa* L.)	(*Triticum aestivum* L.)	Improved plant survival rate, fresh weight, chlorophyll content, higher proline, and soluble sugar contents,
Late embry-ogenesis abundant (HVA1)	Rice (*Oryza sativa* L.)	Barley (*Hordeum vulgare* L.)	Higher leaf relative water content, less reduction in plant growth, better cell membrane protection and drought tolerance

Salt tolerant

Salt Tolerance of Plants

The relative growth of plants in the presence of salinity is termed their salt tolerance. Salt tolerances are usually given in terms of the stage of plant growth over a range of electrical conductivity (EC) levels. Electrical conductivity is the ability of a solution to transmit an electrical current. Salinity is described in terms of electrical conductivity (EC) and measured as ds/m. The higher the EC value, the more is the salinity of soil or water (Table 14).

Table 14. Classification of salinity

Class	ECe (dS/m)	Comments
Non-saline	<2	Salinity has no significant effect
Slightly saline	2-4	Yields of very sensitive crops may be affected
Moderately saline	4-8	Yields of many crops affected
Very saline	8-16	Only tolerant crops yield satisfactorily
Highly saline	>16	Only a few very tolerant crops yield satisfactorily.

The examples of enhanced tolerance to salinity stress through GM approach is furnished below (Table 15).

Table 15. Examples of enhanced tolerance to salinity stress GM approach

Species	Genes transferred	Traits improved
Arabidopsis thaliana	Apoplastic invertase, *Apo-Inv*	Salinity tolerance, Germination increased, Growth
Brassica napus	Arginine decarboxylase, ADC	Salinity tolerance, Germination increased, Yield,
Melon (C*ucumis melo*)	Ca2+-dependent protein kinase, CDPK	Salinity tolerance, , Yield, Growth
Tomato *(Solanum lycopersicum)*	Ca/H antiporter, CAX1	Salinity tolerance, Yield, Growth
Tobacco(*Nicotiana tabaccum*)	Serine/threonine kinase, AT-DBF2	Salinity tolerance, Growth
Rice *(Oryza sativa*)	Calcium-binding protein, EhCaBP	Salinity tolerance, Germination increased,
Potato(*Solanum tuberosum*)	Heat shock protein, DnaK/HSP70	Salinity tolerance, Yield, Growth

Submergence / Flooding / Anoxia tolèrent

When partially or completely submerged, most rice varieties display a moderate capacity to elongate leaves and portion of stems are trapped underwater. This elongation leads to a spindly plant that easily lodges when flood waters recede. If the flood is deep, underwater elongation growth can exhaust energy reserves, causing death within a matter of days. The marker-assisted introgression of the *sub1* region has successfully improved submergence tolerance in a wide range of mega-varieties without any penalties on development, yield, and grain quality. These new lines endure submergence, as long as the flood occurs after the seedling stage but before flowering and the flood completely subsides within 10 to 20 days, depending on floodwater conditions (Table 16). The yield advantage provided by *Sub1* introgression lines is anticipated to greatly stabilize production in rainfed lowland environments that experience flash flooding.

Development of the submergence - tolerant Swarna-*Sub1*

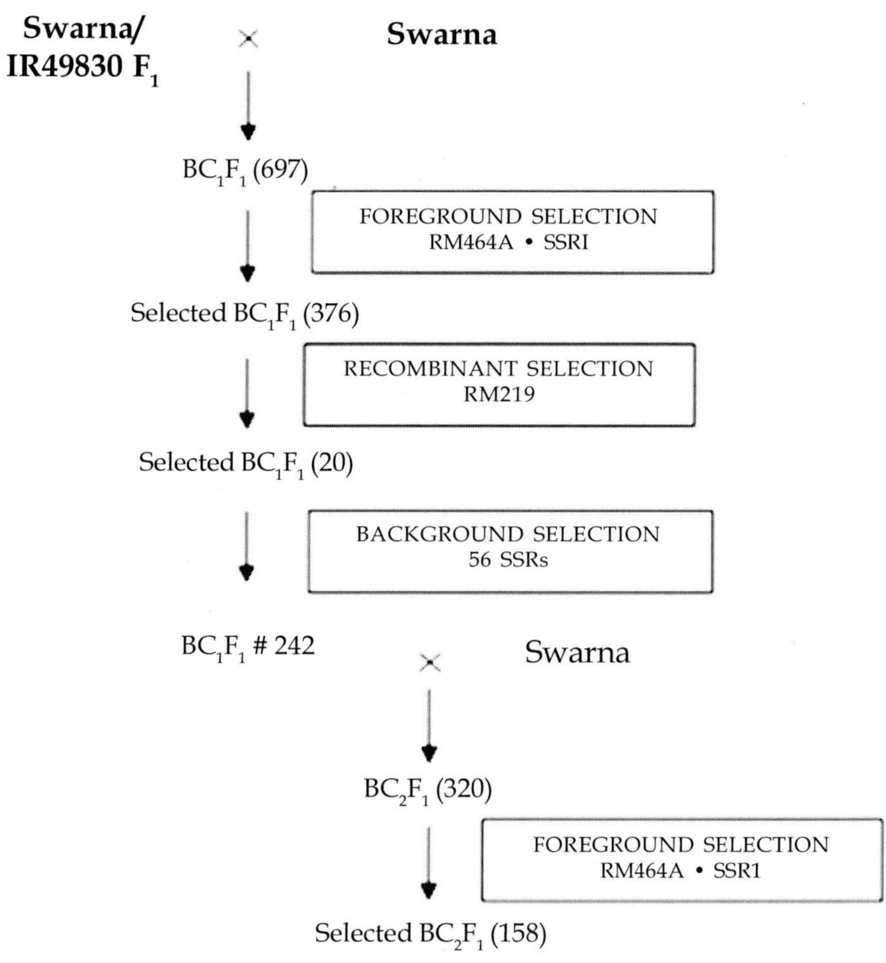

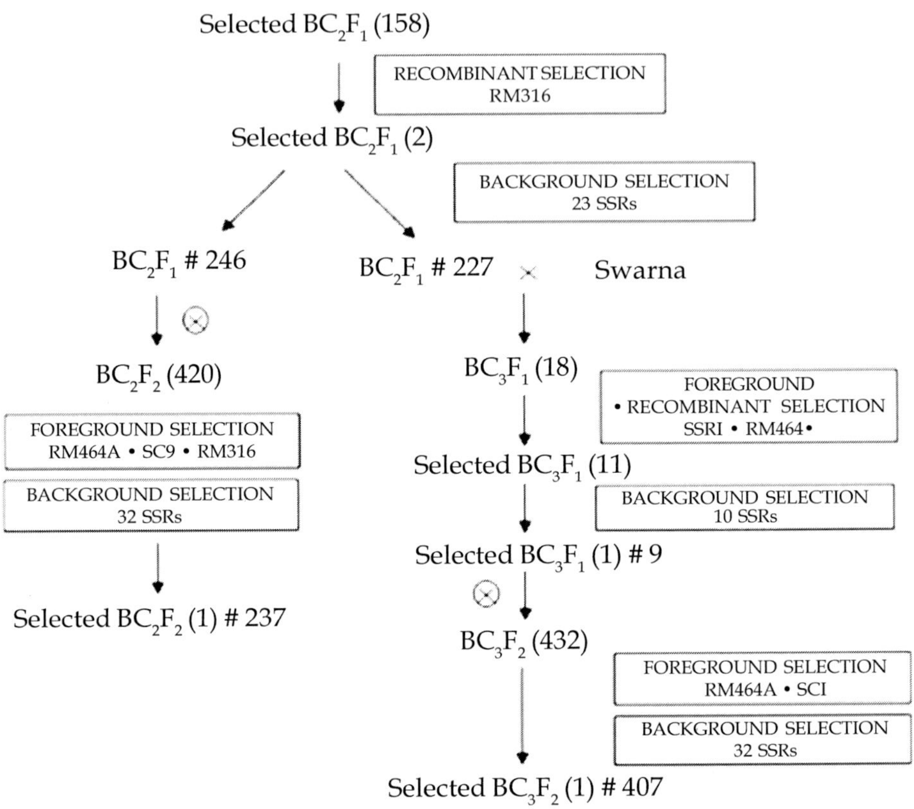

Table 16. Examples of submergence tolerance MAS approaches

Variety or line	QTL	Trait(s) improved
Swarna	*Sub1*	Submergence tolerance
FR-13 A	*Sub1*	Submergence tolerance
IR49830-7	*Sub1*	Submergence tolerance

High Temperature (HT) tolerant

Plants are adapted to limited range of temperatures. If the temperature drops below 15^{0}C, plants experience low temperature stress and if it is above 45^{0}C, plants are subjected to high temperature stress. Manipulation of the heat stock protein response has the potential to improve common stress tolerance that may lead to a more efficient exploitation of the inherent genetic potential of agriculturally important plants (Table 17).

- *Arabidopsis* plants engineered to overproduce glycine betaine, showed increased thermo tolerance, suggesting that nitrogenous osmoprotectants can have a role in cellular protection against heat stress.
- Engineering tobacco plants to express bacterial aspartate decarboxylase, the enzyme catalyzing the decarboxylation of aspartate to beta alanine is another approach.

Table 17. Examples for Heat tolerance

Gene	Source	Plant targeted	Tolerance
HSP 101	*Arabidopsis*	Rice	Heat tolerance
HSP17.7	Carrot	Carrot	Heat tolerance
GmHSFA1	Soybean	Soybean	Heat tolerance
mHSP22	Maize	*Arabidopsis*	Heat tolerance

QTL mapping and genes for low temperature stress tolerance

Freezing Tolerance

Plants that have freeze tolerance will harden as winter approaches, undergoing changes that are necessary for surviving ice growth within the tissues. These changes must occur before the temperature gets too low so that the metabolic demands of tissue remodelling can be met. The alterations that commonly occur consist of an increase in cytoplasmic solutes to act as both, a cryoprotectant as well as to buffer any freeze- concentration of other solutes that could be toxic in high concentrations, as well as membrane alterations that increase the fluidity of membranes at low temperatures and increase the stability of the membranes to low water contents. Sugars possess cryoprotective effect. Sucrose is the predominant one and is closely associated with tolerance in some plant species. Sugar molecules such as fructans, raffinose, sorbitol and mannitol also serve the same functions.

Antifreeze proteins (AFPs)

AFPs are Proteins that decrease the temperature at which ice is formed. They do not affect melting point of solution. AFPs bind to surface of ice nuclei inhibiting the crystal growth. AFPs affect the shape of the ice-crystals, that form inhibitors of recrystallisation (inhibit coalescing of small ice crystals into large ice crystals). AFPs accumulate in apoplastic fluid during cold acclimation. In winter rye, six AFPs (16-35kDa) accumulate in apoplastic fluids. AFPs are endochitinase like proteins having both chitinase and antifreeze activity. Engineering of this antifreeze protein is to mitigate the physical damage caused by ice enhance freezing tolerance.

❏❏❏

Chapter 24

Carbon Sequestration in Soil

Carbon is a key ingredient in soil organic matter (57% by weight). Plants produce organic compounds by using sunlight energy and combining carbon dioxide from the atmosphere with water from the soil. Soil organic matter is created by the cycling of these organic compounds in plants, animals and microorganisms into the soil. Well-decomposed organic matter forms humus, a dark brown, porous, spongy material that provides a carbon and energy source for soil microbes and plants. When soils are tilled, organic matter previously protected from microbial action is decomposed rapidly because of changes in water, air, and temperature conditions, and the breakdown of soil aggregates accelerate erosion. A soil with high organic matter is more productive than the same soil where much of the organic matter has been "burned" through tillage and poor management practices and transported by surface runoff and erosion. However, organic matter can be restored to about 60 to 70% of natural levels with best farming practices.

Carbon sequestration

Carbon is one of the several constituents of soil organic matter. The sequestration of carbon in soils involve the availability of the other building blocks. Carbon sequestration refers to the provision of long-term storage of carbon in the terrestrial biosphere, underground, or the oceans so that the build up of carbon dioxide (the principal greenhouse gas) concentration in the atmosphere will reduce or slow. The sequestering of carbon in soil organic matter involves the production of complex organic structures. Besides carbon, these structures contain hydrogen, oxygen, nitrogen, phosphorus and sulphur. It is a process to increase the soil carbon which could not only reduce the effects of global atmospheric pollution by carbon dioxide, but also to improve the prospects for a long-term solution to the world (country) food problem.

Carbon is found in all living organisms and is the major building block for life on earth. Carbon exists in many forms, predominately as plant biomass, soil organic matter and as the gas CO_2 in the atmosphere and dissolved in seawater. 'Carbon sequestration' implies transferring atmospheric carbon dioxide into long lived pools (oceans, soils, vegetation (especially forests) and geologic formations) and store it securely, so that it is not immediately remitted. Although oceans store most of the earth's C, soils contain approximately 75% of the C pool on land-three times more than the amount stored in living plants and animals. Therefore, soils play a major role in maintaining a balanced global C cycle.

Carbon sequestration in soils

Through the process of photosynthesis, plants assimilate C and return some of it to the atmosphere through respiration. The C that remains as plant tissue is then consumed by animals or added to the soil as litter, when plants die and decompose. The primary way that C is stored in the soil is as Soil Organic Matter (SOM). SOM is a complex mixture of C compounds, consisting of decomposed plant and animal tissues, microbes (protozoa, nematodes, fungi and bacteria), and C associated with soil minerals. Carbon can remain stored in soils for millennia or be quickly released back into the atmosphere. Climatic conditions, natural vegetation, soil texture and drainage affect the amount and length of time, C is stored. Soil is an important reservoir and sink of carbon indeed. The terrestrial carbon (C) pool is the third largest among the five global C pools. The largest or the oceanic pool, contains 38,000 Pg (Pg = petagrams= $1x10^{15}g$ = 1 billion metric tonnes), followed by the geologic pool (comprised of fossil fuel containing 5000 Pg of which 4000 Pg of coal, 500 Pg of oil and 500 Pg of gas), the soil/ pedologic pool containing 2300 pg to one metre depth (comprised of 1550 Pg of soil organic carbon and 750 Pg of soil inorganic carbon), the atmospheric pool containing 760 Pg and finally the biotic pool containing 600 Pg of live mass and detritus material.

Carbon sequestration takes place from the four main components

1. Terrestrial

Carbon dioxide is removed from the atmosphere by photosynthesis and stored in the vegetation, litter and soil organic matter. Enhancing this processes require fundamental understanding of biological and ecological processes and the formation of soil organic matter in both managed and unmanaged terrestrial ecosystems. Identifying ways to enhance carbon sequestration of the terrestrial biosphere include CO_2 removal from the atmosphere by vegetation and storage in biomass and soils.

2. Oceanic

Oceans represent large potential sinks for sequestering CO_2 emissions. Carbon sequestration research goal in oceans is to allow an objective evaluation of direct injection of CO_2 into the midwater or deep ocean and enhancement of natural ocean carbon sequestration by phytoplankton fertilization with nutrients. Enhancing the net oceanic uptake from the atmosphere is by fertilization of phytoplankton with nutrients, and injecting CO_2 to oceans depths greater than 1000 meters.

3. Microbial

Microbes are responsible for much natural CO_2 absorption on land and sea and are also important sources for producing fuels such as methane and hydrogen. Sequencing the genomes of microbes that produce fuels such as methane and hydrogen or aid in carbon sequestration, to allow an evaluation of their potential use to produce, for example, methane or hydrogen from either fossil fuels or other carbonaceous sources, including biomass or even some waste products.

4. Geologic

Geological formations such as brine aquifers, oil and gas fields, and coal beds offer potentially significant large scale opportunities for storage of CO_2. A thorough understanding of geological parameters and processes is needed so that the storage capacity can be maximized and leakages can be minimized. Geosciences research related to understanding the geophysics and geochemistry of potential reservoirs require appropriate attention for subsurface sequestration of carbon dioxide.

Sequestering carbon

Absorbing CO_2 from air and injecting it into the biomass is the only practical way of removing large volumes of GHG from the atmosphere. This would involve a combination of the following set of approaches.

- Reducing deforestation
- Increasing afforestation
- Converting vegetation into timber products which store carbon or substitute bioenergy for fossil fuels so that less carbon is emitted into the atmosphere
- Building up non-living carbon reservoirs such as soils

Potential CO_2 storage reservoirs

Ocean and land-based sites together contain an enormous capacity for storage of CO_2. (Table 18). The world's oceans have by far the largest capacity for carbon storage. (Worldwide total anthropogenic carbon emissions are ~7 GtC per year (1 GtC = 1 billion metric tonnes of carbon equivalent).

Table 18. Capacity for storage of CO_2

Sequestration Option	Worldwide capacity (GtC)
Ocean	1000s
Deep saline formations	100s-1000s
Depleted oil and gas reservoirs	100s
Coal seams	10s - 100s
Terrestrial	10s
Utilization	<10s

Terrestrial carbon sequestration

- Processes through which CO_2 from the atmosphere is absorbed by trees, plants and crops through photosynthesis

- Carbon is stored in biomass (tree trunks, branches, foliage and roots) and soils
- The term "sinks" is also used to refer to forests, crop lands, and grazing lands, and their ability to sequester carbon. Agriculture and forestry activities release CO_2 to the atmosphere
- A carbon sink occurs when carbon sequestration is greater than carbon releases over some time period.

Fossil carbon and non fossil carbon

The total carbon in atmosphere is 760 giga T. Non fossil (vegetation +soil) is 2300 +/- 350 giga T. Fossil carbon (combustible fossils) - 5000 giga T. The total carbon is 8000 giga T.

Agricultural and forestry sequestration

Forests and soils have a large influence on atmospheric levels of carbon dioxide. Tropical deforestation is responsible for about 20% of the world's annual CO_2 emissions. Forests store more carbon than farmland, but farmland has the potential to increase its carbon storage.

Agricultural and forestry practices sequester carbon

1. Avoiding emissions by maintaining existing carbon storage in trees and soils.
2. Increasing carbon storage, e.g., tree planting
3. Substituting bio-based fuels and products for fossil fuels.

Potential carbon sequestration from agricultural and forestry practices

Carbon can be sequestered in the mineral soil after the conversion of intensively cropped agricultural fields to more extensive land uses such as afforested and natural succession ecosystems. Carbon sequestration rates vary by tree species, soil type, regional climate, topography and management practices. Pine plantations can accumulate almost 100 metric tons of carbon per acre after 90 years, or roughly one metric ton of carbon per acre per year (Birdsey, 1996). Changes in forest management (e.g., lengthening the harvest-regeneration cycle) generally result in less carbon sequestration on a per acre basis. Changes in cropping practices, such as from conventional to conservation tillage, have been shown to sequester about 0.1-0.3 metric tons of carbon per acre per year. Young trees store more carbon on an annual basis (kg C/Year) because they are typically growing more rapidly than older trees. Two options for reducing the amount of carbon in the atmosphere are to increase the amount of land planted with permanent grassland or forest vegetation and to reduce the frequency or intensity of tillage operations. Either option would store - or sequester - additional carbon on the affected lands.

Carbon Sequestration in Agriculture

Certain farming practices can reduce GHG emissions and sequester carbon dioxide (CO_2) from the atmosphere. Harnessing methane for energy prevents its escape to the atmosphere, where it acts as a GHG, and displaces the need for other energy sources, which may be derived from fossil fuels. Agricultural activities (e.g., deforestation, burning, ploughing, and intensive grazing) contribute considerably to the atmosphere pool. Expansion of agriculture may have contributed substantially to the atmospheric carbon pool. An important objective of the sustainable management of soil resources is to increase soil organic carbon (SOC) pool by increasing passive or non-labile fraction. Soil surface management, soil water conservation and management, and soil fertility regulation are all important aspects of carbon sequestration in soil.

Conservation tillage, a generic term implying all tillage methods that reduce runoff and soil erosion in comparison with plow-based tillage, is known to increase SOC content of the surface layer. Principal mechanism of carbon sequestration with conservation tillage are increase in micro-aggregation and deep placement of SOC in the sub-soil horizons. Other useful agricultural practices associated with conservation tillage are those that increase biomass production (e.g., soil fertility enhancement, improved crops and species, cover crops and fallowing, improved pastures and deep-rooted crops). It is also relevant to adopt soil and crop management systems that accentuate humification and increase the passive fraction of SOC. Because of the importance of C sequestration, soil quality should be evaluated in terms of its SOC content.

- Agricultural practices can reduce these emissions by increasing the density and depth of carbon in the soil profile and decreasing the decomposition of carbon and losses due to erosion
- Tillage and planting practices that reduce soil erosion and the land application of manure and other organic materials can increase carbon sequestration
- There are also hidden carbon costs associated with fertilizer, pesticides and irrigation due to emissions of CO_2 during manufacturing and from energy consumption
- Changes in agricultural practices have the potential to reduce carbon emissions by 288 million tons/year in the U.S. and 1 billion tonnes/year worldwide.

The carbon sink capacity of the world's agricultural and degraded soils is 50 to 66% of the historic carbon loss of 42 to 78 gigatons of carbon. The rate of soil organic carbon sequestration with adoption of recommended technologies depend on soil texture and structure, rainfall, temperature, farming system, and soil management. Strategies to increase the soil carbon pool include soil restoration and woodland regeneration, no-till farming, cover crops, nutrient management, manuring and sludge application, improved grazing, water conservation and harvesting, efficient irrigation, agro forestry practices, and growing energy crops on spare lands.

An increase of 1 ton of soil carbon pool of degraded cropland soils may increase crop yield by 20 to 40 kilograms per hectare (kg/ha) for wheat, 10 to 20 kg/ha for maize, and 0.5 to 1 kg/ha for cowpea. As well as enhancing food security, carbon sequestration has the potential to offset fossilfuel emissions by 0.4 to 1.2 gigatons of carbon per year, or 5 to 15% of the global fossil-fuel emissions.

Options of Carbon Sequestration in Agricultural Soils

Soil management practices like increasing soil organic carbon content, reduced tillage, manuring, residue incorporation, improving soil biodiversity, micro aggregation, and mulching can play important roles in sequestering carbon in soil (Figure 4). Strategies to sequester carbon in soils also include increase of

i) total soil organic carbon content,

ii) sub-soil organic carbon content,

iii) micro-aggregation and

iv) bio-diversity.

Improved management techniques have shown that scientific agriculture can be a solution to these environmental issues in general and specifically to mitigating the greenhouse effect by increasing soil carbon storage and effectively removing CO_2 from the atmosphere (Figure 5). Many long-term experiments have demonstrated that judicious application of fertilizers with appropriate crop rotations and conservation tillage produce the optimum biomass yield and lead to C sequestration in agricultural soils.

Soil organic carbon in the sub soil can be increased by growing crops with deep root systems for transferring carbon to the sub soil and deep ploughing for incorporation of residue in the sub-soil. Increasing the proportion of soil microaggregates by using long-chain polymers, soil conditioners and earthworms can immobilize large quantities of carbon. Carbon immobilized in such organo-mineral complexes is inaccessible to microorganisms. Besides this, hydrophobic protection helps in limiting the decomposition of fresh organic matter and reduce carbon dioxide emission from agricultural soils. Enhanced sequestration of atmospheric carbon dioxide in the soil, ultimately as stable humus, may prove a lasting solution than sequestering carbon dioxide in the standing biomass through reforestation and aforestation.

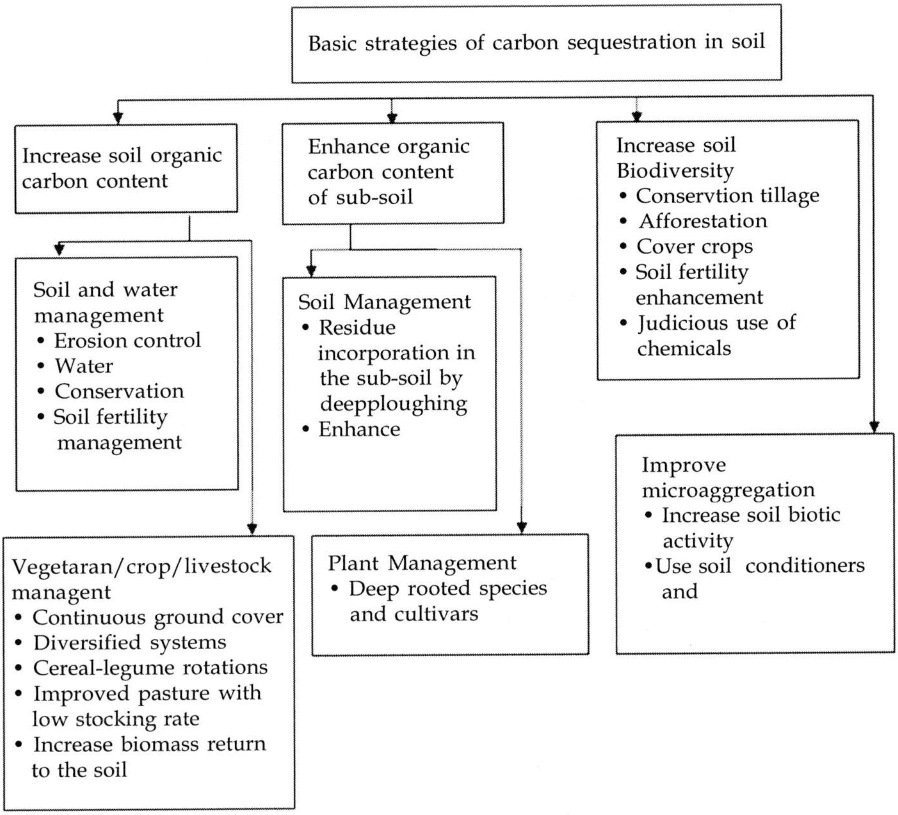

Figure 4: Strategies for carbon sequestration in soils (Lal, 1995)

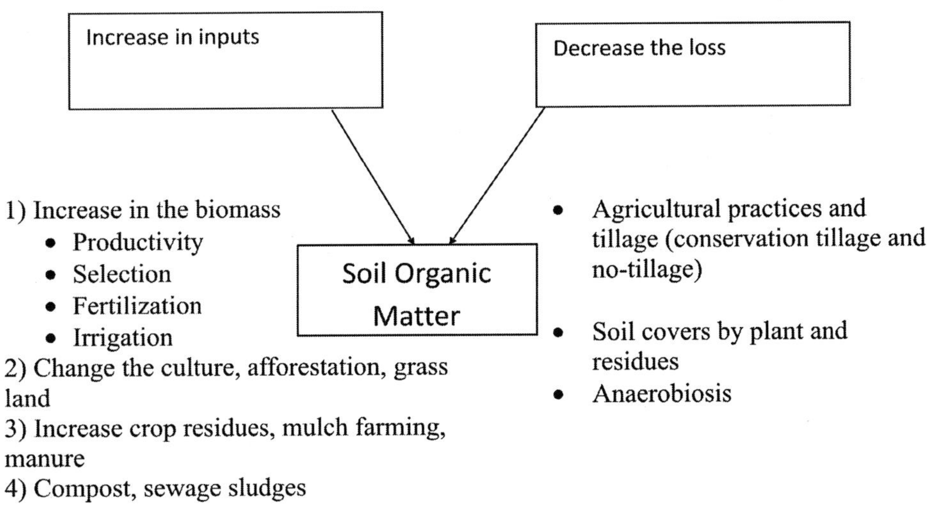

Figure 5. Management of soil organic matter in agriculture

Mechanism of soil carbon stabilization

Other than unfavorable environmental factors, which retard C decomposition in soils (e.g. unfavorable soil pH, temperature, desiccation, anaerobiosis, absence of decomposers, and presence of toxic factors), there is increased possibility for stabilization of C in soils. Kuan meng goh (2004) proposed mechanisms of soil C stabilization could be grouped into physical, chemical and biological protection measures or their combinations.

Physical protection

They largely arose from the interactions of soil organic carbon (SOC) with soil mineral matrix forming strong chemical bonds or causing inaccessibility of the soil C to decompose organisms or their enzymes. According to Larson and Robert (1991), as much as half of the total SOM is maintained and protected in soils by these mechanisms. There are two major groups of mechanisms being responsible for the retention of organic substances (e.g. pesticides) and soil C by clay minerals. These are

1. Physico-chemical stabilization by adsorption of organic matter to clay surfaces (e.g. Cation and anion exchange, polyvalent Cation bridging, H-bonding) forming complexes and
2. Physical stabilization by the penetration of organic matter into interlayer spaces of expanding clay minerals thereby encapsulate and shield the organic matter, inhibiting its accessibility to degrading soil micro-organisms or their enzymes. The adsorption of soil C to clay and silt particles (< 20 ìm) was an important determinant of the stability of SOC in both tropical and temperate soils.

Various studies have showed that tillage caused a loss of SOC due to detrimental direct and indirect effects, attributed to the mechanical (ploughing) and chemical (wetting-drying cycles) disruptions of soil structure, and increased water availability and soil aeration.

Chemical protection

These mechanisms are attributed to the production of charcoal (or black C) by fires and biologically inert or recalcitrant and refractory compounds and very slowly decomposable humic substances (HS) and organic compounds such as lipids (eg. Waxes, cutins, suberins), and chitin by plants, soil fauna and soil micro-organisms.

Biological protection

These include the classical model or aggregate formation and organization in which micro-aggregates are bound together by roots and fungal hyphae, the role of plant root debris and plant-derived rhizodeposits, microbial community diversity in micro-habitats and the formation of regulatory organic compounds by micro biota in the guts of soil arthropods. VAM is one of the beneficial microbes, which can reduce the CO_2 level in the atmosphere without any side effect. It is a heterotrophic endo mycorrhize fungus, which infect the root of the host plant and survive in it by consuming the CHO from the host plant and in exchange, the mycorrhizal plant increase the rate of photosynthesis when compare to non-mycorrizal plants. Hence, the survival of the mycorrhizal plants captured the 30% atmospheric CO_2 and stored in the soil in the form of glomolin. It will act as a finding agent to form aggregate.

Soil Carbon Sequestration for Mitigating Climate Change

Estimates of the total potential of C sequestration in world soils vary widely from a low of 0.4 to 0.6 Gt C/year to a high of 0.6 to 1.2 Gt C / year. Thus, the potential is finite in capacity and time (Patra, 2007)

Agricultural chemicals

Most recommended management practices (RMP) involve C-based input. It takes 0.86 kg C / Kg N, 0.17 kg C / kg P_2O_5, 0.12 kg C / kg k_2O, 0.36 kg C / kg lime, 4.7 kg C / kg of herbicides, 5.2 kg C / kg of fungicides, 4.9 kg C / kg of insecticides, and 150 kg C / ha for pumping ground water for irrigation. Tillage operations emit 15 kg C / ha for mouldboard ploughing, 8 kg C / ha for chisel ploughing and heavy tandem disking, 6 kg C / ha for light tandem disking, 11 kg C / ha for subsoiling, 4 kg C / ha for cultivation and 2 kg C / ha for rotary hoeing. Therefore, conversion from conventional till to no-till farming reduces emission by 30 to 35 kg C / ha per season.

Nutrients

Carbon is only one of the elemental constituents of humus. It is estimated that sequestration of 1 Gt of C in world soils would require 80 million tons (Mt) of N, 20 Mt of P, and 15 Mt of K.

Soil erosion and deposition

Some of the SOC enriched sediments are redistributed over the landscape, others are deposited in depressional sites, and some are carried into the aquatic ecosystems. Although a part of the C translocated by erosion may be buried and redistributed, the rest is emitted into the atmosphere either as CO_2 by mineralization or as CH_4 by methanogenesis. Erosion-induced deposition and burial may be 0.4 to 0.6 Gt C/year compared with perhaps 0.8 to 1.2 Gt C/year emitted into the atmosphere.

Extractive farming practices

Mining SOC from soil for nutrients through organic matter decomposition has an effect on the atmosphere similar to that of fossil-fuel combustion. Therefore, RMPs must enhance rather than deplete SOC pool and soil fertility, increase rather than maintain or decrease crop yield per unit use of fertilizer and other inputs, and improve rather than degrade soil quality.

Hydrologic and carbon cycles

Linking the hydrologic and C cycles through conservation of water resources is crucial to improving agronomic yields and to soil C sequestration in dryland. The low SOC stock in rainfed agriculture can be enhanced through water conservation, water harvesting, and water efficient farming systems. Enhancing SOC stock in dryland ecosystems through no-till farming is important to drought management: a truly win-win option.

Other greenhouse gases

Enhancing SOC stock increases the soil capacity to oxidize CH_4, especially under no-till farming, but may also exacerbate emission of N_2O. Fluxes of CH_4 and N_2O may change the CO_2 mitigation potential of soil management practices and must be considered along with SOC sequestration.

Soils of the tropics

Because of its severe depletion and degradation, the C sink capacity of soils of the tropics may be high, but the rate of sequestration can be low. The need for enhancing soil quality is also more urgent in soils of the tropics than in soils of high latitudes because of low crop yields. Soil restorative farm policies must be implemented to mitigate soil-degradative trends.

Permanence

Soil C sequestration is a natural, cost-effective, and environment-friendly process. Once sequestered, C remains in the soil as long as restorative land use, no-till farming, and other RMPs are followed. Soil sink capacity and permanence are related to clay content and mineralogy, structural stability, landscape position, moisture and temperature regimes, and ability to form and retain stable micro-aggregates.

Benefits of carbon sequestration

Soil carbon or organic matter is important because it impacts all soil quality functions.

- Sustaining biological activity, diversity and productivity. Regulating and partitioning water and solute transport.
- Filtering, buffering, degrading, immobilizing and detoxifying organic and inorganic materials, including industrial and municipal by products and atmospheric deposition, storing and cycling nutrients and other elements within the earth's biosphere.
- The impact of soil organic matter on the soil qualities listed above and soil functions can be summarized as follows:
- Physical effects: soil aggregation, erosion, drainage, aeration, water holding capacity, bulk density, evaporation, and permeability.
- Chemical effects: cation exchange capacity, metal complexing, buffering capacity, supply and availability of N,P,S, and micronutrients, and adsorption of pesticides and other added chemicals.
- Biological effects: activities of bacteria, fungi, actinomycetes, earthworms and other micro organisms are increased.

Constraints to carbon sequestration in soil

Achieving net greenhouse gas emission offsets required change in agricultural operations. Many of the strategies divert land or inputs away from crop or possibly timber production. On the agricultural side, crop prices generally rise while production falls. Exports also are strongly affected. Conversion of crop land to grassland reduces food yield. It must be compensated somewhere else to feed exploding population. So, the farmers will be forced to adopt drastic measures to augment yield from their farmlands. The need to reduce emissions and implementations of emissions trading will also affect fossil feel price. The tax and corresponding transportation costs increase and might influence the cost of petrol based agriculture chemicals and fertilizers as well as price of on and off-farm commodities. Carbon sequestration in the agricultural soils is permanent as long as the farmers use the procedures of conservation agriculture. However, the transition to conservation agriculture is neither spontaneous nor free of cost. There may be extra costs for some tools or equipment. Besides this, shifting to conservation tillage from conventional will result in reduced rate of mineralization, high fertilizer and herbicide requirement and more pollution problems. Leaving crop residues on soil surface will delay the sowing of next crop since a certain time is required for their decomposition. Apart from this, it causes pest and disease infestation. Some practices such as increasing soil fertility via nitrogen fertilization could lead to increased primary productivity and increased carbon storage, but it could also lead to increase in nitrous oxide emission. Also, by increasing soil water content the rates of decomposition will decrease but methane production could increase. As the soil carbon level increases, soil absorption of carbon decrease and soil potential to become a future emission source increases. Subsequent alteration of the management regime can lead to carbon releases. Therefore, adoption of any management practice should be carried out after prior evaluation of its overall effect.

Chapter 25

Climate Change on Nutrient Acquisition

Agriculture is a soil based industry that extracts nutrients from the soil which is necessary to maintain and increase the crop productivity. Nutrients (primary and secondary) are important for plant growth. The absolute or relative absence of any of these nutrients can hamper plant growth; alternatively too high a concentration can be toxic to the plant or to humans. The higher the yield, the greater will be the nutrient requirement.

Tropospheric ozone concentrations have been rising in recent years in many Asian countries. An important root- shoot interaction that might determine the overall response of plants is the ability of the root system to adjust nutrient acquisition capacity to meet variations in shoot demand caused by environmental changes. Plant roots can alter their nutrient acquisition capacity by adjusting their physiological, longevity, morphological and or architectural characteristics to meet changes in shoot nutrient demand. Active root nutrient absorption is a highly adaptive plant characteristic that influences acquisition of N and other nutrients in response to environmental factors.

A shift in biomass allocation to the shoots (a decrease in root-to-shoot ratios, RSR) has been observed in a wide range of plant species in response to O_3 exposure. Reduced biomass allocation to roots may decrease nutrient uptake, as well as reducing the yield of economically important plant parts. Nutrient acquisition is facilitated by increased specific root length and increased fine root numbers.

Nutrient impoverished soils contribute to human malnutrition in two important ways. First, they reduce crop yields, causing food scarcity that results in protein energy malnutrition. Second, crops produced on nutrient poor soils typically have low tissue concentrations of trace elements. Human populations whose diet primarily consists of staple cereal crops (primarily maize, rice, wheat, sorghum, and millet) may meet their protein and energy demands but often suffer micronutrient deficiencies. It is estimated that of the world's human population, 60–80% are Fe deficient, >30% are Zn deficient, 30% are Iodine deficient and about 15% are Se deficient (White and Broadley, 2005). The overwhelming majority of people those suffer from micro- nutrient deficiencies live in developing countries (Kennedy *et al.*, 2003).

If we are somehow able to clear this first hurdle and increase crop yields and nutrient availability by overcoming soil limitations, global climate change also looms large in determining food sufficiency and quality in the 21st century (Rosenzweig and Parry, 1994). Evidence suggests that due to high vulnerabilities and limited resources, developing countries may have limited capacity to implement adaptation measures to achieve food stability in a warmer climate (Mertz *et al.*, 2009). It is well documented that climate warming, and changes in global precipitation patterns, particularly drought, are already affecting crop production in developing countries.

Drought Effects on Nutrient Acquisition

Crop yields on soils in developing countries decrease exponentially with increasing aridity (Lal, 2000). Soil moisture deficit directly impacts crop productivity but also reduces yields through its influence on the availability and transport of soil nutrients. Drought increases vulnerability to nutrient losses from the rooting zone through erosion. Because nutrients are carried to the roots by water, soil moisture deficit decreases nutrient diffusion over short distances and the mass flow of water-soluble nutrients such as nitrate, sulphate, Ca, Mg, and Si over longer distances. Roots extend their length, increase their surface area and alter their architecture in an effort to capture less mobile nutrients such as phosphorus (Lynch and Brown, 2001). Reduction of root growth and impairment of root function under drought conditions thus reduces the nutrient acquisition capacity of root systems.

Drought also disrupts root-microbe associations that are a principal strategy for nutrient capture by plants. Reductions in both carbon and oxygen fluxes and nitrogen accumulation in root nodules under drought conditions inhibit nitrogen fixation in legume crops (Athar and Ashraf, 2009). Drought alters the composition and activity of soil microbial communities which determine the C and N transformations that underlie soil fertility and nutrient cycling (Schimel *et al.*, 2007). For example, soil moisture deficit has been shown to reduce the activity of nitrifying bacteria by slowing diffusion of substrate. Less is known about how drought influences mineralization and decomposition in agricultural systems but it likely slows these processes.

Effects of Intense precipitation on nutrient acquisition

Excessive precipitation can reduce crop yields (Kawano *et al.*, 2009). Intense rainfall events can be a major cause of erosion in sloped cropping systems and where soil instability results from farming practices that have degraded soil structure and integrity. Surface erosion during intense precipitation events is a significant source of soil nutrient loss in developing countries (Zougmore *et al.*, 2009). Because of its high mobility in soil, nitrate leaching following intense rainfall events can also be a significant source of N loss in agriculture (Sun *et al.*, 2008).

Agricultural areas with poorly drained soils or that experience frequent and/or intense rainfall events can have waterlogged soils that become hypoxic. The change in soil redox status under low oxygen can lead to elemental toxicities of Mn, Fe, Al and B that reduce crop yields (Setter *et al.*, 2009), and the production of phytotoxic organic solutes that impair root growth and function. Hypoxia can also result in nutrient deficiency since the active transport of ions into root cells is driven by ATP synthesized through the oxygen dependent mitochondrial electron transport chain (Atwell and Steer, 1990). Significant nitrogen losses can also occur under hypoxic conditions through denitrification as nitrate is used as an alternative electron acceptor by microorganisms in the absence of oxygen (Prade and Trolldenier, 1990).

Effects of temperature on nutrient acquisition

Soil warming can increase nutrient uptake from 100–300% by enlarging the root surface area and increasing rates of nutrient diffusion and water influx. Water soluble nutrients including nitrate, sulphate, Ca, Mg primarily move towards roots through transpiration-driven mass flow (Barber, 1995). Since warmer temperatures increase rates of transpiration, plants tend to acquire water soluble nutrients more readily as temperature increases. Temperature increases in the rhizosphere can also stimulate nutrient acquisition by increasing nutrient uptake via faster ion diffusion rates and increased root metabolism (Bassirirad, 2000). However, any positive effects of warmer temperature on nutrient capture are dependent on adequate soil moisture. If under dry conditions, higher temperatures result in extreme vapor pressure deficits that trigger stomatal closure (reducing the water diffusion pathway in leaves) (Abbate *et al.,* 2004), then nutrient acquisition driven by mass flow will decrease (Cramer *et al.,* 2009). Temperature driven soil moisture deficit slows nutrient acquisition as the diffusion pathway to roots becomes longer as ions travel around expanding soil air pockets.

Emerging evidence suggests that warmer temperatures have the potential to significantly affect nutrient status by altering plant phenology (Nord and Lynch, 2009). The duration of plant developmental stages is extremely sensitive to climate conditions and is particularly responsive to temperature (Cleland *et al.,* 2007). Experimental warming was shown to shorten phenological stages in wheat that resulted in a 9% yield decrease per 1°C increase in temperature (Mitchell *et al.,* 1993). Nord and Lynch (2008) found that genotypes with shorter vegetative growth phases (shortened phenology) had 30% decreases in reproductive tissue and seed production in soil with low phosphorus availability because of reduced P acquisition and utilization. This interaction between warming and P acquisition through shifts in plant phenology like other climate-nutrient interactions likely operate at a global scale.

Because of the important role of SOC in enhancing soil moisture retention, fertility and structure, it has a disproportionately large impact on food security in developing countries (Lal, 2006). Soil organic carbon stocks are the sum of soil organic inputs driven by plant productivity (root exudates, root and shoot turnover) and soil organic losses via heterotrophic respiration and erosion. Warmer temperatures can increase or decrease crop productivity and yield depending on crop type and agricultural zone (Singh *et al.,* 1998). In the tropical and subtropical climates of most developing countries a 2–3°C increase in temperature is expected to diminish crop productivity (Easterling and Apps, 2005). Simulation models predict large losses in agricultural SOC over the 21st century resulting from lower crop productivity (inputs), and higher rates of heterotrophic respiration in response to climate warming (Smith *et al.,* 2009). There is evidence that drier soils under warmer temperatures will also increase SOC losses via higher rates of wind erosion.

Climate warming contributes to the degradation of freshwater quality and availability (IPCC, 2007). A major way in which this happens is through salinization, an important limitation to agriculture productivity in semi-arid regions that is expected to intensify through the 21st century. Increases in demand for irrigation in semi-arid regions such as the Indo-Gangetic plain of northern India and Pakistan and parts of Africa driven by population growth and climate warming are expected to increase the extent of salinization in agriculture. Climate warming can increase agricultural salinization by increasing the demand for irrigation and increasing rates of surface water evaporation. Warmer temperature can also increase salt accumulation in crops via increased transpiration rates. Theoretically, there are reasons to believe that elevated CO_2 may mitigate salt accumulation but empirical evidence does not support that conclusion. In addition, rising sea levels driven by climate warming is expected to contribute to seawater intrusion of coastal aquifers (Antonellini *et al.*, 2008).

Even if the interactions of global climate change and crop nutrition are not well understood, it is probable that the net effects of these changes will be negative for agricultural production in poor nations. Drought induced by higher temperatures and altered rainfall distribution would reduce nutrient acquisition, biological nitrogen fixation, and may disrupt nutrient cycling. More intense precipitation events would reduce crop nutrition by causing short-term root hypoxia, and in the long term by accelerating soil erosion. Increased temperature will reduce soil fertility by increasing soil organic matter decomposition, and may have profound effects on crop nutrition by altering plant phenology. Since soil fertility is already a primary constraint to food security in many developing regions, and crop production is already marginal, these stresses may be disastrous for vulnerable populations. Social adaptation to changing conditions is possible, although most of the technical options face serious obstacles of diffusion and adoption. An urgent effort is required to improve crop nutrition and soil fertility management in poor nations, integrating agro-ecological and socio-cultural aspects of the problem, to avert worsening of a situation that is already desperate.

□□□

Chapter 26

Climate Change on Nutrient Availability

"Soil, and specifically sound soil management, is essential in our continued quest to increase the production of food, feed, fiber, and fuel while maintaining and improving the environment, and mitigating the effects of climate change. Being the essence of all terrestrial life and ecosystem services, we cannot take the soil for granted. Soil is the basis of survival for present and future generations". Soil is a principal component of the global carbon (C) cycle where key interactions between biotic and abiotic components take place to regulate the flow of materials to and from the pedosphere, atmosphere and hydrosphere. Carbon dioxide is removed from the atmosphere and converted to organic carbon through the process of photosynthesis. As organic carbon decomposes, it is converted back to carbon dioxide through the process of respiration. Soils contain about three times more C than vegetation and twice as much as that present in the atmosphere. Soils contain much more C (1500 Pg of C to 1 m depth and 2500 Pg of C to 2m; 1 Pg = 1 x 10 g) than is contained in vegetation (650 Pg of C) and twice as much C as the atmosphere (750 Pg of C).

Carbon in the form of organic matter is a key element to healthy soil. It is estimated that each tonne of soil organic matter releases 3.667 tonnes of CO_2, which is lost into the atmosphere. Similarly, the build-up of each tonne of soil organic matter removes 3.667 tonnes of CO_2 from the atmosphere. Soil organic matter of which Carbon is a major part holds a great proportion of nutrients, cations and trace elements that are of importance to plant growth. It prevents nutrient leaching and is integral to the organic acids that make minerals available to plants. It also buffers soil from strong changes in pH. Organic matter constitutes 1-8% of the weight most soils and nearly all the dry weight of organic soils such as peats. Because of the great weight of soils to the plant rooting depth at which carbon accumulates, the soils of the world store about 1600 Pg of carbon. This represents a carbon storage capacity that is twice that in the atmosphere. In many parts of agricultural soil organic matter content is decreasing at an alarming rate which is categorized as poor soil. Grasslands contain approximately one fifth of the world's global carbon reserves; many of the world's grasslands have been degraded by overgrazing. The content of organic carbon in soils in most cases constitutes less than 5% of the mass of soil material and is generally concentrated mainly in the upper 20 to 40 cm (the so-called topsoil).

Soils are the fundamental foundation of our food security, global economy and environmental quality. Soil quality is largely governed by soil organic matter (SOM) content, which is a dynamic pool and responds effectively to changes in soil management, primarily tillage and carbon inputs resulting from biomass production. Maintaining soil quality and soil health can reduce problems of land degradation, decreasing soil fertility and rapidly declining productivity.

Climate Change on Different Soil Parameters

1. Soil physical parameters

a) Soil water:

Soil water can be fluctuated by a number through climate change such as precipitation causing rapid changes in soil water since the time-scale for response is usually within a few hours, temperature increase resulting in greater evapotranspiration loss of water from the soil and lastly the type of land use. The integral influence of climate-hydrology-vegetation-land use changes are reflected by the field water balance and soil moisture regime.

b) Soil temperature:

There is a close relationship between air temperature and soil temperature and a general increase in air temperature will inevitably lead to an increase in soil temperature. The temperature regime of the soil is governed by gains and losses of radiation at the surface, the process of evaporation, heat conduction through the soil profile and convective transfer via the movement of gas and water.

c) Soil structure and texture differentiation:

Soil structure is an important property which indicates how the soil particles combine together. Soil structure is responsible for the movement of gases, water, pollutants/contaminants, seepage, nutrients, maintenance of water quality, building foundation, soil fauna and the emergence of crops. The nature and quality of the structure is strongly influenced by the amount and quality of organic matter present, inorganic constituents of the soil matrix, cultivation methods and natural physical processes such as shrink-swell (soils with high clay contents, particularly smectitic mineralogy) and freeze-thaw behaviour. A decline in soil organic matter levels lead to a decrease in soil aggregate stability, infiltration rates and increase in susceptibility to compaction, run-off furthermore susceptibility to erosion.

2. Soil biological parameters

1. Soil organic matter:

Soil organic matter is the fraction of the soil that consists of plant or animal tissue in various stages of breakdown (decomposition). Organic matter is made up of different components that can be grouped into three major types: 1. Plant residues and living microbial biomass. 2. Active soil organic matter also referred to as detritus. 3. Stable soil organic matter, often referred to as humus. The living microbial biomass includes the microorganisms responsible for decomposition (breakdown) of both plant residues and active soil organic matter or detritus. Humus is the stable fraction of the soil organic matter that is formed from decomposed plant and animal tissue. It is the final product of decomposition.

The benefits of organic matter can be classified into three groups.

Physical benefits

- Enhances aggregate stability, improving water infiltration and soil aeration, reducing runoff.
- Improves water holding capacity.
- Reduces the stickiness of clay soils making them easier to till.
- Reduces surface crusting, facilitating seedbed preparation.

Chemical benefits

- Increases the soil's CEC or its ability to hold onto and supply over time essential nutrients such as calcium, magnesium and potassium.

- Improves the ability of a soil to resist pH change; this is also known as buffering capacity.
- Accelerates decomposition of soil minerals over time, making the nutrients in the minerals available for plant uptake.

Biological Benefits

- Provides food for the living organisms in the soil.
- Enhances soil microbial biodiversity and activity which can help in the suppression of diseases and pests.
- Enhances pore space through the actions of soil microorganisms. This helps to increase infiltration and reduce runoff.

Soil organic matter is highly susceptible to changes in land use and management, soil temperature and moisture. In the last decades changes in land use and management have already led to a significant decline in organic matter levels in many soils which increases the susceptibility to soil erosion. Climate change caused soil carbon to decrease overall, with a loss of 4 Pg global grasslands after 50 years. Combined climate change and elevated CO_2 increased production and reduced global grassland C losses to 2 Pg, with tropical savannas becoming small sinks for soil C.

Nutrient Availability and Acquisition

Plant availability of nutrients in the soil is a function of soil chemical properties as well as location of the ion relative to the root surface and the length of the pathway the nutrient must travel in the soil to reach the root surface; nutrient acquisition by the plant reflects an array of physiological phenomena that govern nutrient transport to and into roots and can alter aspects of both chemical and positional nutrient availability in the soil. Given that soil moisture and temperature are primary determinants of nutrient availability and root growth and developments and that carbon allocation to roots governs nutrient acquisition, it is reasonable to expect that process outcomes will be reflective of the changed climate. Plants accumulate nutrients from the soil solution pool, and nutrients must be in solution to be mobile in the soil. In the absence of roots, steady-state solution-phase concentrations of nutrient ions are controlled by adsorption-precipitation and desorption-dissolution reactions between nutrients and the surface complex of soil, mineralization and immobilization for solutes of organic origin and additions from fertilizer (Table 19). Given the importance of C and N cycling to both agricultural productivity and sustainability, the preponderance of belowground climate change studies have focused on microbiology. Biological transformation between organic and inorganic pools is strongly influenced by moisture and temperature, and thus, global climate change may strongly influence solution concentrations of N as well as S.

Table 19. Impact of climate change on processes and parameters controlling nutrient availability.

Nutrient availability/ acquisition attribute	Soil/plant controls	Remarks regarding global climate change
Presence of nutrients in soil solution	Adsorption/ desorption Mineralization/ immobilization Fertilization	Increased may increase process rates; temperature (T) increased CO_2 may enhance root exudates that enhance fine root growth and turnover. Increased T may enhance volatilization of surface-applied N fertilizers; changes in rainfall patterns may enhance or depress volatilization and leaching losses of nutrients.
Nutrient movement	Mass flow Diffusion	Increased CO_2 may reduce transpiration, depressing nutrient movement to the root through mass flow but may increase root exudation and fine root growth. Increased T will also enhance nutrient diffusion.
Nutrient uptake	Morphology and architecture Kinetics	Elevated CO_2 may enhance fine root development. If T is suboptimal, increased T will enhance root surface area development; changes in è caused by changes in rainfall patterns may enhance or depress mass flow and/or diffusion.In theory, elevated CO_2 should make more carbohydrate available for active uptake.

Barber (1995).

As a plant transpires, solution-phase nutrients are transported in the convective movement of water in the bulk soil toward root surfaces. Quantitatively, mass flow contributions to a nutrient's acquisition are the product of the volume of water transpired and the mean solution-phase concentrations, mass flow does not deliver sufficient quantities to the root surface. Therefore, in the presence of a growing root, concentrations of these nutrients in the solution-phase concentrations, mass flow does not deliver sufficient quantities to the root surface. Therefore, in the presence of a growing root, concentrations of these nutrients in the solution immediately adjacent to the root surface will be depleted. Movement by diffusion is a function of an ion's diffusivity in water, the water content of the soil, the tortuous nature of the pathway an ion must travel to reach the root, the buffer power and the concentration gradient created by root uptake.

Rising CO_2 concentrations and nutrient availability

- Increased ground biomass - Elevated CO_2 can increase plant biomass, which also stimulates fine root growth and root secretions in soils. Increased root growth generates more carbon below ground, which can help accelerate decomposition and nutrient cycling.
- Increased soil respiration - Increased carbon belowground can increase root respiration rates and soil microbes. This increase in CO_2 from soils limits the amount of carbon stored there. The magnitude of increased respiration varies due to differences in microbial communities, availability of carbon, and plant growth rates.
- Altered decomposition rates - Higher atmospheric concentration of CO_2 can alter the amount of nitrogen in leaf litter, which strongly influences decomposition changes in nitrogen from leaf litter can increase lignin decomposition and decrease organic matter decomposition. These alterations could influence soil carbon availability, alter microbial communities, and change nutrient availability for growing trees and plants.

Rising temperatures and nutrient availability

- Increased soil respiration - Rising temperature increases the rate of chemical and biochemical reactions. For soil microbes and roots, that also means increasing respiration, which can release more carbon dioxide and even methane from soils. Increasing these gases from soils can contribute to existing atmospheric greenhouse gas emissions.

- Loss of stored soil C - Increased decomposition means higher carbon turnover in soils, which could reduce the carbon available for growing plants. While most plants get a majority of their carbon through photosynthesis, soil carbon is important for plant growth and maintenance. Soil carbon helps soil retain water and important nutrients, as well as serves as an energy source for decomposing organisms.
- Higher rates of N mineralization - Increased temperature can increase nitrogen availability through higher turnover of soil nitrogen.

Availability of N, P and K

Successful crop production depends on suitability of climate, edaphic factors and plant type. The characteristics of soil, like soil type, soil mineralogy, texture, etc. and, availability of water are governed by climatic parameters. The total content and plant available fraction of nutrients which strongly determine the growth and yield of crops are also strongly influenced by the climate of particular region because of it is effects on soil formation and development. The microclimate of plants is of particular significance in this regard. The functional activities of plant roots such as absorption of water and nutrients are affected by both the low and high temperatures. The rate of growth and subsequent crop yield depend on prevailing air and soil temperatures. Dark coloured soil absorbs more radiant energy than the light coloured soil. Therefore, dark coloured soils tend to maintain a higher temperature during bright days. Soil types which depends on climate too, affects the nutrient availability. Alkaline soil exhibits the low organic matter and nitrogen deficiency. But in these types of soil the high availability of P and K found. Sometime, alkaline soils have Mo and B in toxic amounts. The rainfall affects the availability of plant nutrients. Soils show the Mn deficiency in the rainy season. However, flooding increases the availability of P because hydrolysis of $FePO_4$ results in the formation of H_2PO_4 which is readily available to the plants. Nutrient shortage in soil is enhanced by drought, excess moisture, unseasonably low temperature, insect damage, disease or mechanical injury.

Nitrogen

Nitrogen is one of the most widely distributed elements in the nature. It is present in the atmosphere, lithosphere and hydrosphere. However, the atmosphere is the main reservoir of N. About 75000 Mg of N is found in the air above one hectare of soil. The atmosphere having 78% of its gaseous content as N_2, is virtually limit less reservoir of this element. This element is very mobile circulating between the atmosphere, soil and living organisms. Nitrogen is found in greater amount in the cooler region, however, its fixation is less due to low temperature, moisture and light intensity. In the hot arid and semi arid regions also nitrogen is deficient. With increases in average annual temperature, organic annual temperature, organic matter content which indicates the status of N, decreases. This decline in organic matter content is sharp with increase in temperature upto 10°C. Climatic parameters and processes govern availability or loses of nitrogen.

Nitrification

The temperature coefficient of nitrogen mineralization is 2, over the range 5 to 35°C. Optimum soil temperature for nitrification of NH_4^+ to NO_3- is 24 to 35°C. N-mineralization and nitrification decreases at soil moisture greater than field capacity. The nitrification rate increases almost linearly with time up to 21 days after application of fertilizer. Nitrification rates are generally highest at soil water content at field capacity or 1/3 bar. N- mineralization and nitrification are reduced when soil moisture exceeds field capacity. Most of the biological reactions are influenced by temperature.

Denitrification

Fertilizer nitrogen enters a labile pool of soil nitrogen which must be subjected to continuous denitrification losses to the atmosphere. The earth's atmosphere is comprised largely of N_2 but its oceans are virtually nitrate free. Denitrification can be useful for removal of excessive amount of NO_3- from irrigation and waste water. The denitrification process is very sensitive to soil temperature, and it increases rapidly in the range of 2 to 25°C. This process proceeds at slightly higher rates when the temperature is increased in the range of 25 to 60°C. However, it is inhibited by temperature greater than 70°C. The rapid increase in denitrification at elevated soil temperature suggests that thermophilic microorganisms have a major role in denitrification.

N-fixation and climate

A high rate of photosynthate production is strongly correlated with increased N-fixation by Rhizobia. Factors that reduce the rate of photosynthesis reduce N-fixation. These factors include reduced light intensity, moisture stress and low temperature. Nitrifying bacteria will not produce NO_3^- in the absence of O_2. Soil conditions that permit rapid exchange of gases into and out of the soil are important for maintaining optimum soil aeration for bacterial activities.

Phosphorus

Phosphorus is absorbed by plans mostly as the primary orthophosphate ion ($H_2PO_4^-$) but some plants, it is also absorbed as secondary orthophosphate. The greater amount of P fixed by 1:1 clays is probably due to higher amount of hydrated oxides of iron and aluminum associated with kaolinite clays. Adsorption of P by iron and aluminium oxides declines with increasing pH. Phosphorus availability in most soils is at maximum in the pH range 6.0 to 6.5. The mobility of phosphate in the soil profile is comparatively low, the uptake of fertilizer P depends much on root growth and the root morphology of the crop.

Effect of temperature

Generally, P adsorption increases with rise in temperature in soils to which fertilizer has been added. P-adsorption in soils of warm region is generally greater in soils of temperate regions. These warmer climates also give rise to soils with higher contents of Fe/Al oxides. Mineralization of P from soil organic matter of crop residue depends on soil biological activity which increases with increasing temperature.

Effect of flooding

In the most soils, there is an increase in available P content after flooding, largely due to conversion of Fe^{3+} - phosphate to soluble Fe^{2+} - phosphate and hydrolysis of Al/Fe phosphate compounds. Other mechanisms of increased P-availability due to flooding include dissolution of occluded P increased mineralization of organic P in acidic soil, increased solubility of Ca-phosphate in calcareous soils and greater diffusion of P. In moderately acidic soils, phosphorus is fixed by hydrous oxides of aluminium andiron and by anion exchange of silicate clays.

□□□

Chapter 27

Climate Change on Ecosystems

The world's climate is changing, and it will continue to change throughout the 21st century and beyond. Rising temperatures, new precipitation patterns, and other changes are already affecting many aspects of human society and the natural world. Climate change is transforming ecosystems on an extraordinary scale, at an extraordinary pace. As each species responds to its changing environment, its interactions with the physical world and the organisms around it change too. This triggers a cascade of impacts throughout the entire ecosystem. These impacts can include expansion of species into new areas, intermingling of formerly non overlapping species, and even species extinctions. Climate change is happening on a global scale, but the ecological impacts are often local and vary from place to place. Human actions have been a primary cause of the climate changes observed today. The Earth's climate is changing as a result of increasing anthropogenic emissions of greenhouse gases (GHGs). Globally, the atmo-sphere and the oceans are warming.

Ecosystem

An ecosystem is a biological environment consisting of all the living organisms or biotic component, in a particular area, and the nonliving, or abiotic component, with which the organisms interact, such as air, soil, water and sunlight. Ecosystems are functional units consisting of living things in a given area, non-living chemical and physical factors of their environment, linked together through nutrient cycle and energy flow. The entire array of organisms inhabiting a particular ecosystem is called a community. The number of species making up such a community may vary from a myriad to a single species such as *Desulforudis*. In a typical ecosystem, plants and other photosynthetic organisms are the producers that provide the food; ecosystems can be permanent or temporary. Ecosystems usually form a number of food webs i.e. net work of food chains.

Types of Ecosystem

Terestial ecosystem-A terrestrial ecosystem is an ecosystem found only on a landform. Four primary terrestrial ecosystems exist: tundra, taiga, temperate deciduous forest, and grassland.

Aquatic ecosystem - An aquatic ecosystem is an ecosystem in a body of water. Communities of organisms that are dependent on each other and on their environment live in aquatic ecosystems. The two main types of aquatic ecosystems are marine ecosystems and freshwater ecosystems.

Lentic ecosystem – Ecosystem of lake, pond or swamp.

Lotic ecosystem – Ecosystem of river, stream or spring.

Biodiversity

Biological diversity, is the diversity of life as per the definition of US Congressional Biodiversity Act, HR1268. Biological diversity means the full range of variety and variability within and among living organisms and the ecological complexes in which they occur, and encompasses ecosystem or community diversity, species diversity, and genetic diversity.

Genetic diversity: is the combination of different genes found within a population of a single species, and the pattern of variation found within different populations of the same species. Coastal populations of Douglas fir (evergreen coniferous trees) are genetically different from Sierran populations. Genetic adaptations to local conditions such as the summer fog along the coast or hot summer days in the Sierra result in genetic differences between the two populations of the same species.

Species diversity: is the variety and abundance of different types of organisms which inhabit an area ie place of settlement.

Ecosystem diversity: encompasses the variety of habitats that occur within a region, or the mosaic of patches found within a landscape.

In the simplest of terms, biological diversity is the variety of life and its processes; and it includes the variety of living organisms, the genetic differences among them, and the communities and ecosystems in which they occur.

Human societies depend on ecosystems for natural, cultural, spiritual, recreational and aesthetic (sensitive) resources they provide. Climate change has potential to alter ecosystems and many resources and services providing to each other and to society. In various regions across world, some high-altitude (eg Alpine biomes) and high-latitude ecosystems were already affected by climate change. A 20 to 30 % of species to be at risk of extinction from climate change impacts within this century if global mean temperatures exceed 2-3 °C relative to pre-industrial levels.These changes cause adverse or beneficial effects on species. For example, climate change benefits certain plant or insect species by increasing their ranges. Their impact on ecosystems and humans, may be positive or negative if invasive (e.g., weeds or mosquitoes) or valuable to humans. (e.g., food crops or pollinating insects)

Animals and plants under increasing threat

Golden toad *(Bufo periglenes)* and harlequin frog *(Atelopus varius)* of Costa Rica (in Central America) disappeared due to global warming, and the species is under threat in more than one way.

Irreversible changes to ecosystems and animals

As climate change wreaks its havoc globally, ecosystems could disappear altogether, or may undergo serious and irreversible changes, such as those happening to coral reefs (**Coral reefs** are underwater structures made from calcium carbonate secreted by corals (marine animals or marine fauna). Coral reefs are colonies of tiny living animals found in marine waters that contain few nutrients.

Warming affects cold seas and polar communities

Polar bears in Hudson Bay of Canada are losing weight and getting less fit because ice breaks up 2 weeks earlier in spring, robbing them of 2 weeks' hunting. The polar bear (*Ursus maritimus*) is a bear native largely within the Arctic Circle (Northern hemisphere) encompassing the Arctic Ocean, its surrounding seas and surrounding land masses. It is the world's largest land carnivore and also the largest bear, together with the omnivorous Kodiak Bear, which is approximately the same size.

Fish stocks used to stay in Cornwall in South England moved as far north as the Shetland Islands. Many animal- or microbe-dominated ecosystems (e.g., marine intertidal zones or coastal bays, ocean and lake bottoms, many streams and rivers and all soils) depend on inputs of organic matter and nutrients from other systems for productivity and biodiversity.

The intertidal zone, also known as the foreshore and seashore and sometimes referred to as the littoral zone, is the area that is above water at low tide and under water at high tide (in other words, the area between tide marks). This area can include many different types of habitats, with many types of animals like starfish, sea urchins, and some species of coral (marine animals). Such systems have critical roles in sustaining ecosystem services such as food production, carbon sequestration, water purification, water flow regulation, and habitat support, due to temperature increases, optimum habitat for many species move to the higher up mountains or further towards the poles, where no higher ground or with quick changes taking place for ecosystems and species to adjust, local losses or even global extinctions will occur.

Marine biological systems under stress

Marine life forms fundamentally well-adapted to natural variations in environment tolerate extreme situations for a limited time. Climate change may alter some habitats severely & stress become too great for many species. When unfavorable factors combine, cumulative effect cause species extinction.

Effects of climate change-related warming of fisheries

Longer growing seasons and increased rates of biological processes are;

- At Greater risk of oxygen depletion.
- Species shift to more tolerant of warmer and perhaps less-oxygenated waters.
- Coastal cultures to consider impacts of sea-level rise on facilities and freeing of contaminants from nearby waste sites.

- Changes in precipitation, freshwater flows and lake levels. Strong regional variations are likely introduce new disease organisms or exotic or undesired species in aquatic ecosystem.
- Establishment of compensating mechanisms or intervention strategies.
- Longer season for production and maintenance.
- Modification of aquaculture systems, e.g. keeping them indoors under controlled light, may be needed for protecting larvae from UVB.

Shifting Penguin populations

- Adélie penguin inhabit winter ice pack & populations decreased 22% during last 25 years.
- Chinstraps (Penguin) remain in close association with open water and populations increased by 400%.
- A 7°-9°F rise in midwinter temperatures on western Antarctic Peninsula in past 50 years, and associated receding sea-ice pack.
- Reflected in changed penguin populations.

Global increase in reactive N inputs (N, NOx and NH_4) to ecosystems. In last decade of 20th century, human-derived N inputs to ecosystems exceeded from all natural processes combined. Presently, accelerated rate of increase in anthropogenic N inputs elevated N inputs alter key metabolic processes (primary production, decomposition & ET).

Changing seasons

- Shorter winters leading to mismatches between key elements in ecosystem, such as feeding periods for young birds & availability of worms or insects for food, also have impact on farmers' growing seasons.
- Long-term records of flowering and nesting times show clear warming trends.In Britain, flowering time and leaf-break (dating back to 1736) provides evidence for climate change.
- Earlier bird breeding, earlier spring migrant arrival & later autumn departure observed in North America, with changes in

migratory patterns in Europe confirm the climate change has already set in.

Effects of elevated CO_2 on seawater chemistry

- Affect silica speciation in seawater, thus affecting siliceous primary producers (diatoms) causing secondary impacts on food webs including fish & marine mammals.
- Affect speciation of metals (Fe, Cd, Co and Mn) and adsorption of metals to suspended particulate matter.
- Resulting into changes in metallic nutrient availability, affect photosynthetic and respiratory processes with complex secondary effects on ecological processes.
- Cumulative changes in water chemistry on biological processes cause feedback to atmosphere by slowing biological CO_2 pump to deep ocean.
- CO_2 changes on calcification rates by marine organisms produce carbonate shells or skeletons have long-term implication.

Coral Reef

Natural balance between reef-building (accretion) and reef-breakdown (dissolution) normally maintains and increases bulk volume and complex architecture of coral reef. Currently the balance is under threat from the following fronts:

- Coral dying, apparently because of more frequent and severe heat waves.
- Detrimental changes in sea-water chemistry caused by increased atmospheric CO_2.
- Mild warming (+2°C) causing tropical near-shore communities change from coral-dominance to algal-dominance.

Coral bleaching

The shallow tropical waters in which most corals are found are warming. Heat stress causes corals to expel the symbiotic algae that provide their primary source of nutrition, leaving only the white "bones" of the corals behind. This process, called coral bleaching, can be lethal to the coral if it lasts too long. Coral bleaching, which has increased in recent decades, becomes worse as high temperatures last longer and longer.

Corals are also being affected by ocean acidification, which is caused by the increase in carbon dioxide in the atmosphere. This affects the ability of marine organisms to build their shells and skeletons. Ocean acidification is likely to slow, or even stop, the growth of coral over this century. This would not only affect the corals themselves but also put in jeopardy the survival of the myriad species found only on coral reefs.

Population and economic growth

Although natural release of greenhouse gases occurs, anthropogenic releases are considered to be the main drivers of climate change. These in turn are influenced by increases in human population, coastal development and economic activity. The link between climate change and economic development is well established. While population growth puts demands on natural resources and leads to development of natural areas, economic development also contributes to the increase in greenhouse gases.

Agricultural Shifts

In general, plants grow faster in warmer climates, which could be good news for some farmers, especially in temperate and cold areas. But this applies only up to a point. When it gets too warm, crops tend to mature too early, and under extreme conditions, high temperatures can kill crops. Although different varieties are bred to withstand certain conditions, each crop ultimately has a limited temperature range.

Dry land ecosystems

- Expanding human populations in arid and semi-arid regions (western United States, Central and Western India) place considerable stress on aquatic and terrestrial ecosystems
- Many dry forest ecosystems in the west are at high risk of uncharacteristic wildfire due to combinations of fire suppression, insect outbreaks and other factors.

Drought and Insects

Climate change can both directly and indirectly affect herbivores and pathogens through various processes. The ultimate effects on patterns of disturbance include increased disturbance in some areas and decreased disturbance in others.

- For example, an increase in the inter annual variation in minimum winter temperatures is expected to favor more northerly outbreaks of southern pine beetles but could reduce more southerly outbreaks.
- Similarly, decreased precipitation and increased evapo transpiration should boost tree secondary chemical metabolism (and, therefore, resistance to pests) in forests that currently suffer modest growing-season water deficits.
- Increased warming would most likely increase the diversity of insects at higher latitudes. Because insects typically migrate much faster than trees, many temperate tree species are likely to encounter nonnative insect herbivores that previously were restricted to subtropical forests.

Insect and pathogen outbreaks

- Climate influences the survival and spread of insects and pathogens directly, as well as the susceptibility of their forest ecosystems.
- Changes in temperature and precipitation affect herbivore and pathogen survival, reproduction, dispersal, and distribution. Indirect consequences of disturbance from herbivores and pathogens include elimination of nesting trees for birds and negative effects on mycorrhizal fungi.
- Other indirect effects include the impacts of climate on competitors and natural enemies that regulate the abundance of potential pests and pathogens.

❑❑❑

Chapter 28

Methods for Estimating Climate Impact on Agriculture

Global climate change has become a major environmental problem affecting the future survival and the development of mankind, and it has attracted widespread attention of governmental organizations and academic community in the world. Agriculture is one of the sectors most sensitive to climate change, and any degree of climate change will bring potential or significant impact to agricultural production and related processes. Climate change has impacted agriculture significantly, and it will inevitably have a huge impact on agricultural production in the future. At present, assessment of climate change vulnerability is still a relatively new field of study. Although IPCC developed a dedicated research program specially for assessment of climate change research, research work thus far indicates feasible methods to evaluate (especially in a quantitative way) different natural systems, human systems and for specific purposes are still lacking.

Since the 1990s, with the deepening of research on climate change impact on agriculture and adaptation the vulnerability of agro-ecosystems and agricultural production to climate change has become the focus of study of scientists around the world. Strengthening the evaluation of agricultural vulnerability to climate change bears more practical significance. By studying the impact of climate change on agriculture and the agricultural vulnerability to climate change in different regions, it is of great significance for proactively adapting to climate change, developing effective adaptation measures and ensuring sustainable agricultural development, and also for providing a scientific basis for decision-making. In addition, scientific and reasonable evaluation of the vulnerability will provide meaningful results for risk analysis of climate change. Therefore, the evaluation of agricultural vulnerability to climate change is not only necessary, but it also is an important part of the research of climate change impact and adaptation.

Agricultural vulnerability to climate change

Definition of agricultural vulnerability

Agricultural vulnerability to climate change has not been clearly defined. At first, there was only qualitative research of food security vulnerability to climate change. The U.S. Department of Agriculture defined vulnerability as the potential for negative consequences that are difficult to alleviate through adaptive measures given the range of possible climate changes that may occur in certain areas of a region, which is the concept of inter-regional comparison.

According to the definition of vulnerability to climate change given by the IPCC assessment report, agriculture vulnerability to climate change is the manifestation of the agricultural sensitivity and adaptive capacity to climate changes, and it changes with the location, time, and socio-economic and environmental situations. Agricultural vulnerability to climate change is the function of characteristics of climate variability, magnitude, and rate of variation within the agricultural system, and the system's sensitivity and adaptive capacity, and it is the degree to which the agricultural system is susceptible to, or unable to cope with adverse effects of climate change including climate variability and extreme events. It is worth noting that in literatures both climate vulnerability and climate change vulnerability have basically the same meaning, which should be differentiated from the vulnerability of agricultural systems themselves.

Different estimation methods

Statistical analysis method

Based on the yield and climate, the variation of climate and yield were analyzed and the vulnerability of autumn grain crops in these areas was estimated. The results showed that the climate represented to become warmer and drier in recent 40 years, which was disadvantageous to productivity increase.

Indicator system method

By use of the results of GCM forecast, the possible changes of disaster indeed are evaluated, the possible change of climate vulnerability of agricultural production in the future in the loss plateau is forecasted.

The sensitivity index was defined as the climate yield divided by climate potential productivity. Adaptation index was the trend yield divided by climate potential productivity. Vulnerability index was equal to sensitivity divided by adaptation. Based on these definitions, crop production vulnerability to climate change was analyzed.

Conclusion

Spatial distribution of vulnerability index was similar to that of the sensitivity index. As for time variation, the magnitude of the increase of the sensitivity indeed was the largest, while the magnitude of the increase of vulnerability indeed was decreasing.

The weight of each index was determined by using analytical hierarchy process, and the method for comprehensively and quantitatively evaluating the vulnerability of agro ecosystem was estimated based on the principles of fuzzy estimate.

Model simulation method

i) By using climate change scenario produced by PRECIS and wheat yield data generated by CERES model, the sensitivity and vulnerability of wheat, maize and rice to future climate change were studied based on the yield variation and GIS mapping.

Conclusion

Preliminary distribution of sensitivity and vulnerability areas of three major grain crops were created

i) Based on drought risk theory of natural vulnerability of crops, natural vulnerability curve of typical wheat varieties was simulated by using the drought hazards strengths of wheat and crop growth module in EPIC.

Quantitatively assessed the spatial and temporal distribution of drought risk of wheat.

ii) To meet integration purpose, it has been proposed a framework for vulnerability assessment based on the clarity of definition in view of local ecological risk ad vulnerability. The framework included eight steps and the indicator structuring technique which uses the vulnerability scoping diagram (VSD) to organize the data in a circular assessment diagram with vulnerability as the centre.

Comprehensive evaluation method

By using designated comprehensive assessment framework, vulnerability to climate change was evaluated, which included devising climate scenarios, analyzing sensitivity, determining indicators of vulnerability, surveying and counting adaption strategies, measuring adaptation capacity, assessing vulnerability and other steps

Estimation models

Stimulation models which integrate current scientific knowledge from many different disciplines, such as crop physiology, plant breeding, agronomy, agrometerology, soil physics, soil chemistry, pathology and entomology, is one of the powerful scientific tools that can be used to assess the impact of climate change impact on different component of agro ecosystem and for evaluate adaptation options.

Crop growth simulation models ad biogeochemical and biophysical models have been very helpful in projecting the future crop and soil productivity scenarios. The models in connection with different GCM models help identifying the future agricultural practices that can adapt to different climate change scenarios. A few models that can be used for different scenarios analysis to combat impact of climate change on agricultural production of the globe have been discussed below.

Decision support system for Agro technology Transfer

DDSAT is a software package integrating the effects of soil, crop phenotype weather and management options that allows users to ask "what if" type questions and simulate results by conducting, in minutes to a desktop computer, experiments which could consume a significant part of an agronomist's career. It has been in use for more than 15 years by researchers in over 100 countries.

The DSSAT simulates growth, development and yield of a crop growing on a uniform area of land under prescribed or simulated management as well as the change in soil, water, carbon and nitrogen that take place under the cropping system over time.

Agro- ecosystem simulation(AGROSIM)

The AGROSIM model family was developed in the ZALF institute for Landscape Modelling in Muncheberg, Germay. The main goal of developing AGROSIM was to have a tool to investigate the consequences of different management strategies on farm level, models of land use and climate change effects on regional level on the basis of computer based simulation runs within whole crop rotations. Till now AGROSIM consists of models for winter wheat (AGROSIM-WW), winter barley (AGROSIM-WG), winter rye (AGROSIM- WR), sugar beet (AGROSIM-ZR) and five different catch crops, which are validated for weather and soil conditions of different locations in northest Germany. All models have a similar structure; need only standard meterological inputs (temperature, radiation, precipitation) generally available soil characteristics and plant parameters.

Cropping Systems Simulation Model

CroSyst is multi-year, multi crop, daily time step crop growth simulation model, developed with emphasis on a friendly user interface, and with a like to GIS software and a weather generator. The model's objective is to serve as a analytical tool to study the effect of cropping system management on crop productivity and the environment. For this purpose, cropsyst simulates the soil water budget, soil-plant nitrogen budget, crop phenology, crop canopy, root growth, biomass production, crop yield, residue production and decomposition, soil erosion by water and pesticide fate. These are affected by weather, management options including crop rotation, cultivar selection, irrigation, nitrogen fertilization, pesticide applications, soil and irrigation water salinity, tillage operation and residue management.

InfoCrop

InfoCrop is a generic crop model, which simulates the effects of weather, soils, agronomic management (planting, nitrogen, residue and irrigation) and major pests on crop growth, yield, soil carbon, nitrogen and water and greenhouse gas emissions. Infocrop considers various processes such as crop-growth and development, soil water, nitrogen and carbon, and crop pest interactions. Each process is described upon the crop/cultivar. The parameters considered in Infocrop are as fallows

- Crop growth and development: phenology, photosynthesis, partitioning, leaf area growth, storage organ numbers, source sink balance, transpiration and uptake, allocation and redistribution of N and water.
- Effects of water, nitrogen, temperature, flooding and frost stress on crop growth and development.
- Crop pest interactions: damage mechanisms of insects and diseases.
- Soil water balance: root water uptake, inter layer movement drainage, evaporation, runoff, ponding.
- Soil nitrogen balance: mineralization, uptake, nitrification, volatilization, inter-layer movement, dentrification, leaching.
- Soil organic carbon dynamics: mineralization and immobilization.

Erosion productivity Impact Calculator (EPIC)

EPIC a cropping systems simulation model, was developed to estimate soil productivity as affected by erosion throughout the United States during the 1980's. EPIC simulates all crops with one growth model using unique parameter values for each crop. The processes simulated include leaf interception of solar radiation; conversion to biomass; division of biomass into roots, above ground mass, economic yield; root growth; water use; and nutrient uptake. EPIC is a field scale, daily time step model composed such as erosion, nutrient balance, crop growth, and related processes. It is designed to stimulate drainage areas that are characterized by homogenous weather, soil, landscape, crop rotation, and management. Since the initial development, EPIC has been continually improving through the additions of algothirms to simulate water quality, climate change and the effect of atmospheric CO_2 concentration, and nitrogen and carbon cycling.

Soil, Water, Atmosphere and Plant (SWAP) model

The model employs the Richards equation including root water extraction to simulate soil moisture movement in variably saturated soils. Concepts are added to account for macro porous flow and water repellency. SWAP considers the basic processes for solute transport such as convection, dispersion, adsorption and decomposition. For more extensive studies which for instance include volatilization or nutrient transformations, SWAP generates soil water fluxes for detailed chemical transport models as PEARL for pesticides and ANIMO for nutrients. SWAP simulates soil heat flow taking into account actual heat capacities and thermal conductivities. The generic crop growth module WOFOST is incorporated to simulate leaf photosynthesis and crop growth. The soil moisture, heat and solute modules exchange status information at each time step to account for all kind of interactions. Crop growth is affected by the actual soil moisture and salinity status on a daily basis. The model is very flexible with regard to input data at the top and bottom of the soil column. Depending on simulation complexity, computation time for 50 year period ranges from 30 to 500 seconds on ordinary personal computers. The SWAP model is suitable to select viable water management options, to perform regional studies employing geographical information systems, to illustrate transport processes for education and extension. In recent years, SWAP has been employed to explore alternative flow and transport concepts, to analyze laboratory and field experiments, and to evaluate management options with respect to field scale water ad solute movement.

- Field scale water and salinity management
- Irrigation scheduling
- Transient drainage conditions
- Plant growth affected by water ad salinity
- Pesticide leaching to ground water and surface water
- Regional drainage from top soils towards different surface water systems
- Optimatization of surface water management
- Effects of soil heterogenicity

Denitrification – Decomposition (DNDC)

DNDC is a computer simulation model of C and N biogeochemistry in agro ecosystems. The model can be used for predicting crop growth, soil temperature and moisture regime, soil carbon dynamics, nitrogen leaching, and emission of trace gases including nitrous oxide, nitric oxide, ammonia, methane and carbon-di-oxide(CO_2).

The DNDC model is 1- D process based biogeochemical model and consists principally of two components. The first component includes the sub models for soil, climate, decomposition and plant growth. The second component includes the sub-models for nitrification and denitrification. The concentrations and fluxes of NO_3, N_2O, NO and N_2 are calculated based o simulated soil microbial activities, which depend on simulated soil environment conditions and a series of biochemical and geochemical reactions determining the transport and transformation of C and N components to allow simultaneous occurrence of nitrification and denitrification in an aerobic or anerobic microsites.

Agricultural Production Systems SIMulator (APSIM)

APSIM was developed to simulate biophysical processes in farming systems, particularly as it relates to the economic and ecological outcomes of management practice. APSIM is structured around plant, soil and management modules. These modules include a diverse range of crops, pastures ad trees, soil pH, erosion and a full range of management controls.

The APSIM modeling framework is made up of the following components:

- A set of biophysical modules that stimulate biological and physical processes in farming systems
- A set of management modules that allow the user to specify the intended management rules that characterize the scenario being simulated and that controls the simulation.
- Various modules to facilitate data inputs and output to and from the simulation.
- A simulation engine that drives the simulation process and facilitates communication between the independent modules

- Various user interfaces and association database tools for visualization and future analysis of output
- A web based user and developer support facility that provides documentation, distribution and defect/change request tacking

Models in climate change research

Several studies have used crop models to study the impact or sensitivity of climate change on agricultural production. The effects of climate change and elevated CO_2 concentration on wheat production using APSIM model in Australia has been investigated. These include the assessment on individual effects of rising temperature and CO_2 concentration and rainfall changes and compound effects of temperature, CO_2 and rainfall changes on multiple probabilistic climate change scenarios. However, impact of climate change differed significantly between soil types and locations, and in the higher rainfall region of south Western Australia, what yield is likely to increase yield especially at drier site while higher temperature had a positive effect in the cooler and wetter southern part of the region. The main difference between soil types was that heavier clay soils are most vulnerable to reduced rainfall while sandy soils were more vulnerable to higher temperatures. Higher temperature could both increase and decrease protein concentrations.

Ludwig and Asseng (2010) used cop soil simulation model to test changes in physiological traits, related to early vigour and flowering time, effects on increased yield when compared to traditional cultivars of wheat grown at higher temperatures, elevated atmospheric CO_2 and lower rainfall in a Mediterranean climate. They reported through simulation studies that increase rooting depth generally had a positive impact on yield, while lower radiation use efficiency and earlier flowering tended to reduce yield. The interaction between these traits was generally positive and the results indicated that early vigour may improve yield for a range of future climate scenarios.

DNDC model is a process-based model that originally focused on N_2O and CO_2 emission. However, it has expended to stimulate NO, CH_4, and NH_3 emissions. Using this model, environmental impacts such as land use type, agricultural activities can be assessed in a comprehensive way. The model has been applied to estimate N_2O emissions from agricultural fields and dairy farms.

The DSSAT model is decision support system that encompasses process-based computer model that predict growth, development and yield as a function of local weather and soil conditions, crop management scenarios ad genetic information. It can be used at a farm level as well as regional level to determine the impact of climate change, production and potential adaptation practices that should be developed for farmers. The DSSAT crop model projected the shorter vegetative and reproductive growing season for maize and winter wheat during 21st century. Projected yield reduction for winter wheat varied between 0 and 7 % during the 2020s and 2050s, and between 4 and 20 % in the 2080s.

The CENTURY ecosystem model is used to investigate the effect of land use and climate on soil organic matter (SOM) and plant growth. CroSyst simulates the soil budgets, soil plant nitrogen budget, crop phenology, canopy and root growth, biomass production, crop yield, residue production and decomposition, soil erosion by water and salinity. CroSyst model to predict barley yield reduced by 4 and 5 % under changing climatic conditions. Tubellio *et al.* (2000) used CropSyst model for quantifying potential effects of future climate on agricultural production for four different cropping systems and suggested that combined effects of elevated atmospheric CO_2 and climate change depress crop yields. Predicted warmer air temperatures accelerated plant phenology, reduced dry matter accumulation and crop yields by 10 -40%. For irrigated maize and soybean production, irrigation water requirement increased by 60-90% under climate change scenarios to maintain grain yields.

InfoCrop provides integrated assessment of the effect of weather, variety, pests, and management practices on crop growth and yield, as well as on soil nitrogen and organic carbon dynamics in aerobic as well as anaerobic conditions and greenhouse gas emissions. Climate change is expected to influence crop production, hydrological balances, input supplies and other components of agricultural systems. However, the nature of these biophysical effects and human responses to them are complex and uncertain. Agricultural systems are also dynamic; producers and consumers are continuously responding to changes in crop yields, food process, input prices, resource availability and technological change. Accounting for these adaptations and adjustments is difficult but necessary in order to measure accurately climate change impacts. It is important to estimate crop yield by the use of crop biophysical simulation models that embed parameters drawn from crop experiments. Because climate change is likely to cut across a host of environmental factors, most qualitative estimates of climate change effects on crop yields are derived from such crop simulation models. While the use of crop simulation models makes tractable assessment of climate effects across a range of crops, such models also have limitations including variability of factors and conditions that affect production in the field.

The research prospects of agricultural vulnerability to climate change

Although the assessment study of agricultural vulnerability to climate change made some findings, research methods and tools are also under rapid development, so far no widely accepted method has been found. Major problems in the existing studies and then directions of the development are as follows.

1. Currently, the indicator system method was widely used to study vulnerability problems, because of the complex relationship among affecting factors and multiple factors involved and other reasons, how to scientifically construct a reasonable indicator system and determine the index become the bottleneck of the indicator system method. More research work in building target, determining weights and mutual relations of indices is needed in future.
2. The adaptation capacity is an important aspect of vulnerability. Quantitative research of adaptation capacity is still scarce at present, which should be strengthened in future.
3. The analysis and assessment of vulnerability are two aspects of vulnerability research. Vulnerability of agricultural systems and agricultural vulnerability to climate change should be scientifically identified and distinguished in research, and the problem of uncertainty should be considered in the evaluation process.
4. In the assessment of future agricultural vulnerability to climate change, lack of socio-economic scenario data and relatively simple climate scenarios currently applied need to be studied. Besides, attention should be paid to availability and reliability of data and information. With deepening research of climate change, the methods and tools for vulnerability assessment will be

innovated and developed. The focus of the research will be moving from the natural ecosystem to a coupled system (human-environment coupling system, social-ecological system, human-land system, etc.). The scale and level of evaluation will continue to change, and the uncertainties in researches will be constantly identified and reduced.

Economic approaches to measuring climate change effects

A number of economic approaches and models are found in existing economic assessments. A simple taxonomy of these methods is to classify them as either 'structural' or 'spatial-analogue' approaches. The characteristics of each approach are described here.A structural approach is interdisciplinary, using models from several disciplines to measure economic consequences of climate change. This general approach thus directly incorporates the effects of climate change on yield. These crop simulation models can be adjusted to include the direct effects of increased atmospheric CO_2, and specific farm-level adaptation activities such as planting dates, switching crops, and adding irrigation. The number of included adaptations, such as adding or enhancing irrigation, changing crop mixes, treatment of crop 'migration' potential, or changing input mixes, varies across studies.

After measuring crop yield changes under different climates (e.g. from GCM forecasts), the yield estimates can then be incorporated into economic models of the agricultural sector to estimate changes in acreage and supply by crop and by region and consequent changes in market clearing prices. The economic models seek to either minimize costs or maximize consumer and producer welfare, subject to the climatic and other constraints imposed on the model. This approach has been applied at the state, regional, and national level.

Spatial-analogue models can use either statistical or programming methods to analyze changes in spatial patterns of production; however, both methods assume costless, structural adjustment and adaptation. A statistical approach to analyze cross-sectional data of current agricultural production across both warmer and cooler regions. They examined the relationship between agricultural land values and climate using county-level data in the United States.

□□□

Chapter 29

Long Term and Short Term Projections of Climate Change Effects on Natural Vegetations

Geographic variations in temperature, rainfall and seasonality determine ecosystem productivity and species diversity. Ecosystems therefore respond to changes in temperature and precipitation, which inevitably have an impact on biodiversity. Recent shifts in the distributions of various species towards the poles and to higher altitudes and the extinction of more than 1% of all amphibian species indicate that climate change is already having a major impact on biodiversity. Climatic changes are also expected to alter the distributions of most types of vegetation and there is already evidence of a shift from deciduous woodland to evergreen forest in part of southern Europe. Such changes will have implications both for biodiversity and for the humans that rely on natural ecosystems to survive. Climate is a key factor in determining where individual species of plants and animals can live, grow and reproduce.

Vegetation change is one of the most important results of a changing climate because it affects the habitats of most terrestrial species. Dynamic global vegetation models (DGVMs) have been developed to simulate how various climates and atmospheric carbon dioxide (CO_2) concentrations could affect vegetation around the world. Simulation model showed that by 2045, the most extensive changes are likely to occur in the parts of the world that are already dry - such as central Asia - and along the existing boundaries of major vegetation types. The model predicts increases in the amount of vegetation and woody cover in many dry areas because an increase in atmospheric CO_2 will allow plants to survive increased levels of drought. Furthermore, the boundaries of major vegetation types will shift along moisture and temperature gradients. Overall, vegetation change will disproportionately affect countries with high biodiversity.

The proportion of land surface expected to change from one vegetation class to another within each country is positively correlated with the total number of species and the number of threatened species currently within those countries, but is not significantly correlated with the number of endemic (confined to particular place or country or locality) species. Similarly, the average woody plant cover change predicted by 2045 is positively correlated with the total number of species, the number of endemic species and the number of threatened species currently present. Hence, most vegetation change is predicted to take place in countries that currently support high levels of biodiversity. Although the absolute magnitude of the projected temperature change is larger at high latitudes, the relatively high annual variation in the climate at high latitudes means that the species there have already experienced and survived a wide range of conditions around the historical mean, including occasional warm years

- The impacts of Climate change will be felt disproportionately by countries with relatively low fossil fuel-based CO_2 emissions and low GDPs.
- The effects of climate change are apparent within ecosystems around the world, including medicinal and aromatic plant populations.
- Medicinal and Aromatic Plants (MAP) in Arctic and alpine areas face challenges associated with their rapidly changing environments, and some researchers have raised concerns regarding the possible losses of local plant populations and genetic diversity in those areas. The potential loss of MAP species from effects of climate change is likely to have major ramifications i.e., consequence or development on the livelihoods of large numbers of vulnerable populations across the world.
- Flavonoids are also common in Arctic plants, and they protect against UV damage such as cell apoptosis (a pattern of cell death) related to DNA breakage. If Arctic plants begin to produce less of these compounds as a result of higher temperatures, they may partly lose their ability to serve as antioxidants for human health benefit.

- Higher air temperatures, mainly since the end of the 1980s, have led to advanced timing of phonological phases and consequently to an extension of the general growing season. Phonological models developed to estimate the impact of climate change on the natural vegetation are based on a regional climate scenario for the specific state where such information is required.
- The results indicate that changes in the timing of phenophases could continue in the future, due to distinct temperature changes in winter and in summer, mainly the spring and summer phases will be advanced. Spring phenophases, such as leafing or flowering, show the strongest trends. Depending on the species, the average timing of these phenophases could be advanced by 3-27 days by 2050. The annual growth period of individual trees will be further extended, mainly because of the shift of spring phases.
- Frequent droughts in summer and in autumn can compensate for the earlier leafing of trees, when climatic factors such as temperature and precipitation change in a region beyond the tolerance of a species.

Phenotypic plasticity

Phenotypic plasticity is the ability of an organism to change its phenotype in response to changes in the environment. Phenotypic plasticity, the capacity of a single genotype to exhibit variable phenotypes in different environments, is common in insects and is often highly adaptive. All plasticity is physiological, but can manifest as changes in biochemistry, physiology, morphology, behaviour, or life history. Phenotypic plasticity can be passive, anticipatory, instantaneous, delayed, continuous, discrete, permanent, reversible, beneficial, harmful, adaptive or non-adaptive, and generational. Virtually any abiotic or biotic factor can serve to induce plasticity, and resulting changes vary from harmful susceptibilities to highly integrated and adaptive alternative phenotypes, numerous physiological mechanisms accomplish plasticity, including transcription, translation, enzyme, and hormonal regulation, producing local or systemic responses. Plasticity is important because it is an encompassing model to understand life on earth, it can increase fitness, generate novelty, and facilitate evolution, it structures ecological communities, and it has numerous practical applications. As such, all biologists should understand phenotypic plasticity.

Effect of climate change on biodiversity

Changes in climatic factors such as temperature and rainfall may directly affect the distribution, lifecycles, habitat use, physiology and extinction rates of individual species. In turn these changes may modify the structure and composition of certain ecosystems and communities, by altering competition and other interactions between species. In altered ecosystems, invasive species are likely to thrive, while rare species may become extinct.

Climate change may also indirectly affect species and ecosystems by altering important factors such as:

- fire frequency and behaviour
- the degree of dry land salinity
- the spread of diseases such as dieback
- water flows in rivers and wetlands
- the level of groundwater
- the frequency of extreme climatic events (floods, hail, tropical cyclones, drought)
- ocean acidity levels

These factors along with existing environmental stresses may greatly reduce the ability of our biodiversity to adapt to climate change naturally. Visible signs of the impact of climate change on our landscape may include disappearing wetlands, coral death due to coral bleaching and greater coastal erosion. Visible impacts of climate change on our biodiversity include changes in the flowering times of native plants, changes to the arrival and departure times of some bird species, a movement away from the usual locations by native animals, a decline in fish species and a greater spread of weeds and pests.

Uncertainty of climate change on biodiversity

Because ecological systems are very complex and future climate conditions are not easy to predict, it is difficult to confidently project the impacts of climate change on our biodiversity. Modelling climate change impacts on biodiversity requires good information about a large number of factors viz.

- Distribution of species
- Role of climate upon the distribution of species
- Degree to which species can naturally adapt to climate change
- Capacity of species to migrate
- Potential influence of climate change on threatening processes
- Influence of higher atmospheric concentrations of carbon dioxide on plant and animal physiology
- Ecological processes and interactions between species
- Evolutionary history of species and previous responses to natural variations in climate.

Assisting species to adapt to the impacts of climate change: In many cases, humans may need to intervene to help natural systems to adapt to the impacts of climate change. These interventions include:

Protecting or establishing habitat corridors: Corridors allow plants and an animal to migrate through the landscape as climate changes alter environmental conditions and habitats.

Protecting existing and identifying new, refuges: Refuges are areas that provide natural sanctuaries for plants and animals. Refuges can protect small populations, geographically restricted populations or remaining populations. Populations that occur in past climatic refuges will also protect increased genetic diversity that may allow adaptation to future climatic conditions. Identifying the characteristics of future refuges is an important activity as it influences the design of the State's national parks and conservation reserves.

Building resilience (recovery) to climate change: One way to protect species and ecosystems from the impacts of climate change is to reduce the effect of existing threatening processes, particularly those that may be affected by climate change.

Safeguarding the most vulnerable species: Vulnerable species are those with long generation times, low mobility, highly specific host relationships, small or isolated areas in which they can live and/ or low genetic variation. Where climate change is likely to result in such species becoming locally does not exist, seed collection and storage or protection in a garden or zoo will be required. Protecting species in this way is an essential element of a comprehensive biodiversity conservation program, and will be an increasingly important option as climate change proceeds in this century.

Natural Vegetation

- Natural vegetation in India primarily comprised of forests
- *Vegetation* growing in correspondence with different environmental conditions - Natural Vegetation of particular place.
- Several major factors (soil, topography, temperature and RF) influencing Natural Vegetation of India to large extent.

Features characterizing Natural Vegetation of India are:

- Tropical deciduous forests
- Tropical rain forests
- Alpine and tundra vegetation
- Rain forests of Southern India
- Himalayan vegetation
- Desert region
- Temperature forests and
- Grasslands and many more.

Global natural vegetation distribution depends greatly on climate, climate change due to increasing GHGs in the atmosphere affect distribution pattern of world's natural vegetation. CO_2 amounts to about half of the GHGs and contributes subsequently to GH effect, forests contain about 90 % of carbon stored in terrestrial vegetation. Deforestation currently occurring at a very rapid rate, resulting in decreased carbon storage capacity in forests, necessary to evaluate climate change effects on global natural vegetation distribution for addressing importance of forest management to conserve global environment & sustainable production of forest resources.

Climate Change Effects on Crops

- More rapid maturation of grain and oilseed crops

- Greater risk of crop failures, if precipitation decreases or more variable
- More rapid growth of weeds, northward migration of weed species
- Greater resistance of weeds to herbicide applications
- Invasion by exotic grass species into arid lands causing increased fire frequency
- Increased growing season length by 10 to 14 days compared to last 19 years across temperate latitudes i.e., northern latitudes

Potential climate change impact on Indian forests

- In Western Ghats: >2000 M (MSL) with wide tropical vegetation, with temperature increase & reduced frost, montane (dense growth of trees) forests expand into grasslands
- In Central India, increased south west monsoon replacing drier Sal to moister Teak in drier belt
- In North-East India, increased temperatre, shifting of lower altitude tropical and sub-tropical forests to higher altitude temperate forests occur

Climate Change Effect in Montane Region

- Montane (dense forest) regions (>2000 m MSL) of Western Ghats in Southern India feature stunted evergreen forests (C_3 type) interspersed with extensive grasslands (C_3 or C_4 types)
- Shift in C_3 and C_4 plant types related to changes in moisture and atmospheric CO_2, with lower moisture and CO_2 levels favouring the latter plant types (C_4)
- Expansion of montane forest and replacement of C_4 with C_3 grassland expected in future climates
- Human impact on natural vegetation, such as conversion of grasslands to monoculture plantations of wattle and Eucalyptus interfere with natural succession caused by global climate change

Climate Change, Fire & Natural Vegetation

- Significant climate change affecting frequency & severity for ignition and fire spread
- Extended growing season & lack of summer precipitation aggravating length & severity of forest fire
- Lightening being major source of fire, wind velocity determining fire spread speed And drought frequency and duration cause fire potential

Effect of increased sea level on vegetation

- Rising sea levels causing displacement of most densely populated coastlines in world, threatening freshwater sources and mangrove ecosystems
- Increased frequency and intensity of floods
- Increased vulnerability of people in coastal, arid and semi-arid zones of country
- Over 50% of India's forests likely to experience shift in forest types, adversely impacting associated biodiversity, regional climate dynamics as well as livelihoods based on forest products

Climate & Vegetation

- GHGs affect biochemical processes altering ecological relationships
- Massive species migrated during warmer periods & to higher elevations (eg. U.S.) zones
- Microclimate, soils, life cycle processes (germination & growth) and ecological interactions (competition, herbivory and fire) affect external morphology & physiology of plant communities within vegetation

Policies Addressing Climate Change Impacts

- Incorporate climate concern in long-term forest policy-making process
- Conserve forests and reduce forest fragmentation

- Expand protected areas and link them wherever possible to promote migration
- Promote mixed species forestry to reduce vulnerability
- Undertake anticipatory planting and assist natural migration through transplanting plant species
- Promote *in situ and ex situ gene pool conservation*
- Initiate forest fire management strategies.

❒❒❒

Chapter 30

Carbon Trading

Climate change is expected to hit developing countries the hardest. Its effects viz., higher temperatures, changes in precipitation patterns, rising sea levels, and more frequent weather-related disasters pose risks for agriculture, food, and water supplies. Green house gas (GHG) emissions in the form of methane, carbon dioxide, nitrous oxide, sulphur dioxide adversely affect the global climate by destroying the ozone layer, leading to global warming. It is recognized by world leaders and through Kyoto Protocol a legally binding emission target for the member countries has been proposed. It is a two way concept where by countries can directly reduce emissions with cleaner or greener technologies or absorb emissions by growing forests. Carbon credits significantly reduce greenhouse gases by earmarking a monetary value to the cost of polluting the atmosphere. More importantly, it has the potential of bringing together a number of underdeveloped countries to participate in International efforts to deal with climate change. The greatest value of carbon credits is that it will benefit the rural poor in combating climate change.

Carbon Credits

Carbon Credits are part of a tradable permit scheme. They provide a way to reduce greenhouse gas emissions by giving them a monetary value. A credit gives the owner the right to emit one tonne of carbon dioxide (One tonne of CO_2 should be equivalent to one carbon credit). Credits represent the right to emit a specific amount of greenhouse gases. Credits can be exchanged between businesses or bought and sold in the international market at the current market price. It is a simple, non-compulsory way to counteract the greenhouse gasses that contribute to climate change and global warming. Carbon credits create a market for reducing greenhouse emissions by giving a monetary value to the cost of polluting the air.

The Carbon Credit is the new currency and each carbon credit represents one tonne of carbon dioxide either removed from the atmosphere or saved from being emitted. It is also called emission permit. It is a certificate awarded to countries that are successful in reducing emissions of greenhouse gases. Carbon credits are generated as the result of an additional carbon project. For example, if an environmentalist group plants enough trees to reduce emissions by one ton, the group will be awarded a credit. If a steel producer has an emissions quota of 10 tons, but is expecting to produce 11 tons, it could purchase this carbon credit from the environmental group. The carbon credit system looks to reduce emissions by having countries honor their emission quotas and offer incentives for being below them.

Definition

Carbon Trading is a scheme where firms (or countries) buy and sell carbon permits as part of a programme to reduce carbon emissions. Usually firms are given a certain quota to pollute a certain amount. If they wish to pollute more than their allowance, then they have to buy more permits. If they pollute less than their quota, they can sell their spare permits on the market. Thus, there is an incentive to reduce pollution and find the most efficient way of dealing with pollution. Over time, governments can reduce pollution quotas to encourage greater efficiency.

Examples of Carbon Trading

An early example of an emission trading system has been the SO_2 trading system under the framework of the Acid Rain Program of the 1990 Clean Air Act in the U.S. Under the program, which is essentially a **cap-and-trade emissions trading system,** SO_2 emissions were reduced by 50% from 1980 levels by 2007.

The European Union Emission Trading Scheme (or EU ETS) is the largest multi-national, greenhouse gas emissions trading scheme in the world. It is one of the EU's central policy instruments to meet their cap set in the Kyoto Protocol. This fiscal incentive should encourage the development of technology which leads to more efficient engines and reduces carbon emissions. In theory, it is also socially efficient, i.e. a firm can pollute more if it feels it is worth the cost of buying more carbon permits.

Benefits of Carbon Trading

- The argument is that firms are free to choose the most cost effective way of meeting permit requirement. For example, they have an incentive to develop better technology which limits carbon emissions. However, if the price of permits is low, they may decide to buy more.
- Gradual reductions in Permits. The idea with carbon trading is for governments to gradually reduce the amount of available permits from year to year. Therefore, firms need to increasingly find more ways to reduce carbon emissions.

Arguments against Carbon Trading

- Complex: It can become complicated especially with regard to number of permits to be given. For example when the EU introduced a system of carbon trading, in the initial period of 2005-07, the price of carbon permits were driven down to zero because the EU misjudged the amount of permits. However, any scheme will take a while to be effective.
- Difficulty of measuring how much a firm is actually polluting.
- Transaction costs involved in buying and selling permits.
- Free Rider Problem: The problem of excess carbon emissions is a global problem. Therefore, there needs to be a global solution. If carbon trading is introduced in one country but not others, it may cause production to shift to countries without the scheme. Often countries don't start carbon trading precisely for the fear other countries will be free riding on their efforts.
- Carbon Tax may be more simple, transparent and easy to administer.
- Carbon Trading may have a greater impact on those with low incomes and poor areas who have less flexibility to change their lifestyles.

Carbon Tax – Pros and Cons

The production of carbon dioxide is widely held to contribute to social / environmental problems such as global warming. This carbon pollution is a negative externality. It is a cost imposed on the whole of society and not just the individual who consumes a certain product. E.g. if you drive a car, the external costs are felt by everyone else. Because certain carbon intensive industries create negative externalities, the social cost of production is greater than the private cost. In a free market, these negative externalities are not included in the price leading to overconsumption and social inefficiency. We can say there is a missing market, because the external cost of carbon emission is ignored.

The Purpose of a Carbon tax

The purpose of a carbon tax is to internalise the externality. What this means is that the final price of the goods should include the external cost and not just the private cost. In theory, the tax should equal the external cost. Therefore, the price consumers pay will be the social cost. Demand will fall and the new equilibrium will be socially efficient because at this output, the marginal social cost equals the marginal social benefit.

Revenue Neutral

In theory, a carbon tax should be revenue neutral. This means the tax raised from taxing carbon emissions can be used to reduce other taxes. There should be no overall increase in the tax burden.

Problems of Carbon Tax

- Production may shift to other countries with no or lower carbon taxes.
- Cost of administrating the tax.
- Difficult to know the level of external cost and how much the tax should be.
- Possibility of tax evasion. Higher taxes may encourage firms to hide carbon emissions.
- If demand is price inelastic, the tax may have to be very high to reduce demand significantly.

- Consumers dislike new taxes and often don't believe that they will be 'revenue neutral'. This is not an economic argument, but it is a political reality and explains why it is often difficult to implement.

Market trend

Carbon emissions trading has been steadily increasing in recent years. According to the World Bank's Carbon Finance Unit, 374 million metric tonnes of carbon dioxide equivalent (tCO_2e) were exchanged through projects in 2005, a 240% increase relative to 2004 (110 $mtCO_2e$) which was itself a 41% increase relative to 2003 (78 $mtCO_2e$). In terms of dollars, World Bank has estimated that the size of the carbon market was 11 billion USD in 2005, 30 billion USD in 2006, and 64 billion in 2007.

U.N. Framework Convention on Climate Change (UNFCCC)

The United Nations Framework Convention on Climate Change (UNFCCC) was opened for signature at the 1992 United Nations Conference on Environment and Development (UNCED) in Rio de Janeiro (known by its popular title, the Earth Summit). On June 12, 1992, 154 nations signed the UNFCCC, that upon ratification committed signatories' governments to a voluntary "non-binding aim" to reduce atmospheric concentrations of greenhouse gases with the goal of "preventing dangerous anthropogenic interference with Earth's climate system." These actions were aimed primarily at industrialized countries, with the intention of stabilizing their emissions of greenhouse gases at 1990 levels by the year 2000; and other responsibilities would be incumbent upon all UNFCCC parties. The parties agreed in general that they would recognize "common but differentiated responsibilities," with greater responsibility for reducing greenhouse gas emissions in the near term on the part of developed/industrialized countries, which were listed and identified in Annex I of the UNFCCC and thereafter referred to as "Annex I" countries.

System of emission credits

It refers to a system of credits that allows a company or country that reduces its carbon-dioxide emissions below a target level to sell the extra reduction as a credit to a company or country that has not met the target level. Parties to UNFCCC are classified as:

- Annex I countries - industrialized countries and economies in transition
- Annex II countries - developed countries which pay for costs of developing countries
- Developing countries.

Annex I countries which have ratified the Protocol have committed to reduce their emission levels of greenhouse gasses to targets that are mainly set below their 1990 levels. They may do this by allocating reduced annual allowances to the major operators within their borders. These operators can only exceed their allocations if they buy emission allowances, or offset their excesses through a mechanism that is agreed by all the parties to UNFCCC.

- Annex II countries are a sub-group of the Annex I countries. They comprise the OECD members, excluding those that were economies in transition in 1992.
- Developing countries are not required to reduce emission levels unless developed countries supply enough funding and technology. Setting no immediate restrictions under UNFCCC serves three purposes:
    1. It avoids restrictions on their development, because emissions are strongly linked to industrial capacity
    2. They can sell emission credits to nations whose operators have difficulty in meeting their emissions targets
    3. They get money and technologies for low-carbon investments from Annex II countries.

Developing countries may volunteer to become Annex I countries when they are sufficiently developed.

Some opponents of the Convention argue that the split between Annex I and developing countries is unfair, and that both developing countries and developed countries need to reduce their emissions unilaterally. Some countries claim that their costs of following the Convention requirements will stress their economy. Other countries point to research results that the cost of compliance to be less than the cost of the consequences of doing nothing.

Annex I countries

There are 40 Annex I countries and the European Union is also a member. These countries are classified as industrialized countries and countries in transition: Australia, Austria, Belarus, Belgium, Bulgaria, Canada, Croatia, Czech Republic, Denmark, Estonia, Finland, France, Germany, Greece, Hungary, Iceland, Ireland, Italy, Japan, Latvia, Liechtenstein, Lithuania, Luxembourg, Monaco, Netherlands, New Zealand, Norway, Poland, Portugal, Romania, Russian Federation, Slovakia, Slovenia, Spain, Sweden, Switzerland, Turkey, Ukraine, United Kingdom, United States of America.

Annex II countries

There are 23 Annex II countries and the European Union. Turkey was removed from the Annex II list in 2001 at its request to recognize its economy as a transition economy. These countries are classified as developed countries which pay for costs of developing countries:

Australia, Austria, Belgium, Canada, Denmark, Finland, France, Germany, Greece, Iceland, Ireland, Italy, Japan, Luxembourg, Netherlands, New Zealand, Norway, Portugal, Spain, Sweden, Switzerland, United Kingdom, United States of America

The Marrakesh Accords of the Kyoto protocol defined the international trading mechanisms and registries needed to support trading between countries, with allowance trading now occurring between European countries and Asian countries. However, while the USA as a nation did not ratify the Protocol, many of its states are now developing cap-and-trade systems and are looking at ways to link their emissions trading systems together, nationally and internationally, to seek out the lowest costs and improve liquidity of the market. However, these states also wish to preserve their individual integrity and unique features. For example, in contrast to the other Kyoto-compliant systems, some states propose other types of greenhouse gas sources, different measurement methods, setting a maximum on the price of allowances, or restricting access to CDM projects. Creating instruments that are not truly fungible would introduce instability and make pricing difficult. Various proposals are being investigated to see how these systems might be linked across markets, with the International Carbon Action Partnership (ICAP) as an international body to help co-ordinate this.

Kyoto Protocol

As the Intergovernmental Panel on Climate Change (IPCC) reports came in over the years, they shed abundant light on the true state of global warming and they gave support to the environmental effort to address this unprecedented problem. However, the same discussions that started decades back had never ceased and the crusade for a tangible solution to global climate change had gone on all the while. In 1997, the Kyoto Protocol was adopted. The Kyoto Protocol is a 1997 international treaty which came into force in 2005. In the treaty, most developed nations agreed to legally binding targets for their emissions of the six major greenhouse gases (GHS). Emission quotas (known as "Assigned amounts") were agreed by each participating 'Annex 1' country, with the intention of reducing the overall emissions by 5.2% from their 1990 levels by the end of 2012. The United States is the only industrialized nation under Annex I that has not ratified the treaty, and is therefore not bound by it. The Intergovernmental Panel on Climate Change has projected that the financial effect of compliance through trading within the Kyoto commitment period will be limited at between 0.1-1.1% of GDP among trading countries. The Protocol defines several mechanisms ("flexible mechanisms") that are designed to allow Annex I countries to meet their emission reduction commitments (caps) with reduced economic impact (IPCC, 2007).

Under Article 3.3 of the Kyoto Protocol, Annex 1 Parties may use GHG removals, from afforestation and reforestation (forest sinks) and deforestation (sources) since 1990, to meet their emission reduction commitments.

Annex 1 Parties may also use International Emissions Trading (IET). Under the treaty, for the 5-year compliance period from 2008 until 2012, nations that emit less than their quota will be able to sell Assigned amount units to nations that exceed their quota. It is also possible for Annex I countries to sponsor carbon projects that reduce greenhouse gas emissions in other countries. These projects generate tradable carbon credits that can be used by Annex I countries in meeting their caps. The project-based Kyoto Mechanisms are the Clean Development Mechanism (CDM) and Joint Implementation (JI).

The CDM covers projects taking place in non-Annex I countries, while JI covers projects taking place in Annex I countries. CDM projects are supposed to contribute to sustainable development in developing countries, and also generate "real" and "additional" emission savings, i.e., savings that only occur thanks to the CDM project in question (Carbon Trust, 2009). Whether or not these emission savings are genuine is, however, difficult to prove (World Bank, 2010).

Carbon leakage

Carbon leakage is the effect when regulation of emissions in one country/sector has on the emissions in other countries/sectors that are not subject to the same regulation. There is no consensus over the magnitude of long-term carbon leakage. In the Kyoto Protocol, Annex I countries are subject to caps on emissions, but non-Annex I countries are not. The leakage rate is defined as the increase in CO_2 emissions outside of the countries taking domestic mitigation action, divided by the reduction in emissions of countries taking domestic mitigation action. Accordingly, a leakage rate greater than 100% would mean that domestic actions to reduce emissions had the effect of increasing emissions in other countries to a greater extent, i.e., domestic mitigation action had actually led to an increase in global emissions.

Estimates of leakage rates for action under the Kyoto Protocol ranged from 5 to 20% as a result of a loss in price competitiveness, but these leakage rates were viewed as being very uncertain. For energy-intensive industries, the beneficial effects of Annex I actions through technological development were viewed as possibly being substantial. This beneficial effect, however, had not been reliably quantified.

Emissions trading (also known as **cap and trade**) is a market-based approach used to control pollution by providing economic incentives for achieving reductions in the emissions of pollutants.

A central authority (usually a governmental body) sets a limit or *cap* on the amount of a pollutant that can be emitted. The limit or cap is allocated or sold to firms in the form of emissions permits which represent the right to emit or discharge a specific volume of the specified pollutant. Firms are required to hold a number of permits (or *carbon credits*) equivalent to their emissions. The total number of permits cannot exceed the cap, limiting total emissions to that level. Firms that need to increase their emission permits must buy permits from those who require fewer permits. The transfer of permits is referred to as a trade. In effect, the buyer is paying a charge for polluting, while the seller is being rewarded for having reduced emissions. Thus, in theory, those who can reduce emissions most cheaply will do so, achieving the pollution reduction at the lowest cost to society.

Carbon Trading in Selected Countries

1. Australia

There are active trading programs in several air pollutants. For greenhouse gases, the largest is the European Union Emission Trading Scheme. In the United States, there is a national market to reduce acid rain and several regional markets in nitrogen oxides. Markets for other pollutants tend to be smaller and more localized.

In 2003, the New South Wales (NSW) state government unilaterally established the NSW Greenhouse Gas Abatement Scheme to reduce emissions by requiring electricity generators and large consumers to purchase NSW Greenhouse Abatement Certificates (NGACs). This has prompted the rollout of free energy-efficient compact fluorescent light bulbs and other energy-efficiency measures, funded by the credits. This scheme has been criticised by the Centre for Energy and Environmental Markets (CEEM) of the UNSW because of its lack of effectiveness in reducing emissions, its lack of transparency and its lack of verification of the additionality of emission reductions.

2. European Union

The European Union Emission Trading Scheme (or EU ETS) is the largest multi-national, greenhouse gas emissions trading scheme in the world. It is one of the EU's central policy instruments to meet their cap set in the Kyoto Protocol (Jones *et al.*, 2007).

After voluntary trials in the UK and Denmark, Phase I commenced operation in January 2005 with all 15 (now 25 of the 27) member states of the European Union participating. The programme caps the amount of carbon dioxide that can be emitted from large installations with a net heat supply in excess of 20 MW, such as power plants and carbon intensive factories and covers almost half (46%) of the EU's Carbon Dioxide emissions. Phase I permits participants to trade amongst themselves and in validated credits from the developing world through Kyoto's Clean Development Mechanism.

A number of design flaws have limited the effectiveness of scheme (Jones *et al.*, 2007). In the initial 2005-07 period, emission caps were not tight enough to drive a significant reduction in emissions (CCC,

2008). The total allocation of allowances turned out to exceed actual emissions. This drove the carbon price down to zero in 2007. This oversupply reflects the difficulty in predicting future emissions which is necessary in setting a cap.

Phase II saw some tightening, but the use of JI and CDM offsets was allowed, with the result that no reductions in the EU will be required to meet the Phase II cap (CCC, 2008). For Phase II, the cap is expected to result in an emissions reduction in 2010 of about 2.4% compared to expected emissions without the cap (business-as-usual emissions) (Jones *et al.*, 2007). For Phase III (2013–20), the European Commission has proposed a number of changes, including:

- the setting up of an overall EU cap, with allowances then allocated to EU members;
- tighter limits on the use of offsets;
- unlimiting banking of allowances between Phases II and III;
- and a move from allowances to auctioning.

In January 2008, Norway, Iceland, and Lichtenstein, joined the European Union Emissions Trading System (EU ETS) according to a publication from the European Commission. The Norwegian Ministry of the Environment has also released its draft National Allocation Plan which provides a carbon cap-and-trade of 15 million metric tonnes of CO_2, 8 million of which are set to be auctioned. According to the OECD Economic Survey of Norway 2010, the nation "has announced a target for 2008-12, 10% below its commitment under the Kyoto Protocol and a 30% cut compared with 1990 by 2020."

3. Tokyo, Japan

The Japanese city of Tokyo is like a country in its own right in terms of its energy consumption and GDP. Tokyo consumes as much energy as "entire countries in Northern Europe, and its production matches the GNP of the world's 16th largest country". Originally, Japan had its own cap and trade system that had been in place for some years, but was not effective. Japan has its own emission reduction policy but not a nationwide cap and trade program. This climate strategy is enforced and overseen by the Tokyo Metropolitan Government (TMG). The first phase, which is alike to Japan's scheme, runs up to 2014, these organizations will have to cut their carbon emissions by 6%; those who fail to operate within their emission caps will from 2011 on be required to purchase emission allowances to cover any excess emissions, or alternatively, invest in renewable energy certificates or offset credits issued by smaller businesses or branch offices. Firms that fail to comply will face fines. According to local reports, organizations that do not operate within their caps will also be ordered to cut emissions by 1.3 times the amount they failed to reduce during the first phase of the scheme. The long term aim is to cut the metropolis' carbon emissions by 25% from 2000 levels by 2020.

4. United States

An early example of an emission trading system has been the SO_2 trading system under the framework of the Acid Rain Program of the 1990 Clean Air Act in the U.S. Under the program, which is essentially a cap-and-trade emissions trading system, SO_2 emissions were reduced by 50% from 1980 levels by 2007. Some experts argue that the cap-and-trade system of SO_2 emissions reduction has reduced the cost of controlling acid rain by as much as 80% versus source-by-source reduction. In 1997, the State of Illinois adopted a trading program for volatile organic compounds in most of the Chicago area, called the Emissions Reduction Market System. Beginning in 2000, over 100 major sources of pollution in eight Illinois counties began trading pollution credits.

In 2003, New York State proposed and attained commitments from nine Northeast states to form a cap-and-trade carbon dioxide emissions programme for power generators, called the Regional Greenhouse Gas Initiative (RGGI). This programme was launched on January 1, 2009 with the aim to reduce the carbon "budget" of each state's electricity generation sector to 10% below their 2009 allowances by 2018. Also in 2003, U.S. corporations were able to trade CO_2 emission allowances on the Chicago Climate Exchange under a voluntary scheme. In August 2007, the Exchange announced a mechanism to create emission offsets for projects within the United States that cleanly destroy ozone-depleting substances.

Also in 2003, the Environmental Protection Agency (EPA) began to administer the NOx Budget Trading Program (NBP) under the NOx State Implementation Plan (also known as the "NOx SIP Call") The NOx Budget Trading Programme was a market-based cap and trade program created to reduce emissions of nitrogen oxides (NOx) from power plants and other large combustion sources in the eastern United States. NOx is a prime ingredient in the formation of ground-level ozone (smog), a pervasive air pollution problem in many areas of the eastern United States. The NBP was designed to reduce NOx emissions during the warm summer months, referred to as the ozone season, when ground-level ozone concentrations are highest. In March 2008, EPA again strengthened the 8-hour ozone standard to 0.075 parts per million (ppm) from its previous 0.008 ppm.

In 2006, the California Legislature passed the California Global Warming Solutions Act, AB-32, which was signed into law by Governor Arnold Schwarzenegger. Thus far, flexible mechanisms in the form of project based offsets have been suggested for five main project types. A carbon project would create offsets by showing that it has reduced carbon dioxide and equivalent gases. The project types include: manure management, forestry, building energy, SF_6, and landfill gas capture.

Since February 2007, seven U.S. states and four Canadian provinces have joined together to create the Western Climate Initiative (WCI), a regional greenhouse gas emissions trading system. July 2010, a meeting took place to further outline the cap-and-trade system which if accepted would curb greenhouse gas emissions by January 2012.

On November 17, 2008 President-elect Barack Obama clarified, in a talk recorded for YouTube, his intentions for the US to enter a cap-and-trade system to limit global warming. The 2010 United States federal budget proposes to support clean energy development with a 10-year investment of US $15 billion per year, generated from the sale of greenhouse gas (GHG) emissions credits. Under the proposed cap-and-trade programme all GHG emissions credits would be auctioned off, generating an estimated $78.7 billion in additional revenue in FY 2012, steadily increasing to $83 billion by FY 2019.

Renewable Energy Certificates

Renewable Energy Certificates, or "green tags", are transferable rights for renewable energy within some American states. A renewable energy provider gets issued one green tag for each 1,000 kWh of energy it produces. The energy is sold into the electrical grid, and the certificates can be sold on the open market for profit. They are purchased by firms or individuals in order to identify a portion of their energy with renewable sources and are voluntary. They are typically used like an offsetting scheme or to show corporate responsibility, although their issuance is unregulated, with no national registry to ensure there is no double-counting. However, it is one way that an organization could purchase its energy from a local provider who uses fossil fuels, but back it with a certificate that supports a specific wind or hydro power project.

Global Brand

Carbonfund organisation has established America's first and the leading global brand for products that have no net harm on our climate. Companies that earn the Carbon*Free*® Certified label have reported:

- Increased sales and profits
- Greater customer satisfaction and loyalty
- Enhanced brand value and marketing opportunities
- Carbon finance facilitates the financial reward through carbon credits for the reduction of greenhouse gas emissions by emitters in developing countries. Activities of the World Bank Carbon Finance Unit (CFU) are part of the larger global effort to combat climate change, and are aligned with the World Bank and its Environment Department's mission to reduce poverty and improve living standards in the developing world.

Mexico

The Mexican Housing Authority estimates a deficit today of approximately 8.9 million homes — a number that increases annually by 200,000. With a total of 24.3 million homes in Mexico as of the 2005 census, the country needs more than 35% additional homes for the current population. The deficit is expected to increase over the next two decades, while the population is projected to grow from 37.8 million people between the ages of 25 and 45 (the prime home-buying age group) in 2005 to 45.8 million by 2030, resulting in a dramatic upsurge in the number of potential buyers. In addition to housing deficit concerns, Mexico is already suffering from the effects of year-round climate change in terms of higher temperatures and lower rainfall.

Mexico must move from a carbon-intensive to a carbon-neutral economy. At present, the country emits 715 million tons of carbon dioxide into the atmosphere each year. Total greenhouse gas emissions exceed the per capita emissions in Brazil and India and on par with those of China. Left unchecked, the consequences of climate change and water scarcity are expected to have adverse effects on social and financial conditions in Mexico.

Sustainable housing can help reduce these problems. In the context of Mexico, sustainable housing refers to the homes and communities developed with the objective of (a) reducing environmental impact through the use of ecological materials, equipment and practices; (b) improving the quality of life for society by creating a social fabric that fosters prosperous societies, as opposed to simply building commuter cities with little to no interaction among inhabitants, and (c) increasing access to financing for those in need while promoting savings through the efficient use of water and energy.

Thailand

BANGKOK (AlertNet) - Carbon trading, in which the right to emit the greenhouse gas is bought and sold, is considered an important tool in the fight against climate change. But a growing group of researchers and environmental activists say taxing carbon offers a fairer, more transparent and less corruptible way of limiting emissions. A carbon-added tax (CAT) could be applied to imports, placing the financial burden on the end users of products. In contrast to carbon trading, it would spread the burden wider, as well as preventing companies from lobbying for free carbon credits and market speculators profiting from price fluctuations.

Governance Risks

Carbon trading has been promoted by the developed world as one of the few practical and politically acceptable solutions to tackle climate change. According to Toru Kubo, Head of the Asian Development Bank's carbon market programme, the real amount of money flowing into projects that reduce greenhouse gas emissions is in the order of several billion dollars.

Underlining the governance risks, emission permits were recently stolen from accounts in the Czech Republic and Austria, leading the European Union to impose a week-long freeze in spot trading on its 72 billion euro carbon market during January 2011. EU officials have promised action to improve security and re-open the system as each national registry proves it can resist further attacks. But the incident, the latest in a string of scandals including the re-sale of used offsets, VAT fraud and hacking, will add to factors discouraging other countries from adopting similar cap and trade schemes. The United States delayed last year plans to launch its own system, mainly due to slow progress on domestic legislation, leading to similar delays in Australia and Japan.

Carbon trading in India

India and China are the members of CDM board under UN Framework Convention for Climate Change. These two countries are said to have reaped most of the benefits from carbon trading.

India will be unveiling its own cap and trade mechanism to deal with air pollution on February 1, 2011. The proposed market-based mechanism seeks to introduce a system of self-regulation among industrial units by putting a price on emission of pollutants. A price for emissions makes it costly to pollute, therefore incentivising polluters to reduce emissions.

This is a shift from the command control system, where state pollution control boards determined the emission levels of individual units to a market-based system, where a price is set on emissions. The regulator or state pollution control board sets an overall limit for emissions of different pollutants and industrial units self-regulate to ensure that these limits are not breached. Broadly, the state pollution control board sets a limit on the amount of categories of air pollutants that can be emitted on the basis of its desired concentration in the atmosphere. The state regulator then allocates through permits the "acceptable" level of emissions to industrial units. The industrial units can trade this right to emit, so units which exceed the set level will have to buy permits from those who manage to restrain emissions to that below the cap. This will help in lowering pollution levels at lower overall costs of compliance.

There are several pre-requisites of this scheme that will have to be addressed. These include reliability in data monitoring, estimation of accurate baselines and strong regulatory framework. Basically a transparent mechanism, with robust monitoring and institutional preparedness, is a must for the scheme to take off. The state regulators will need to consider how to implement the scheme in the unorganised sector where monitoring may not be easy.

The pilot project has been devised by a team from MIT/Harvard which includes economist Esther Duflo. It draws on the US experience of dealing with acid rain—reducing sulphur dioxide emissions—in the mid-nineties, the emissions offsets trading programme for total suspended particulate matter in Chile in the early nineties, and the emissions trading scheme for carbon dioxide in the European Union.

India and World Bank

India and the World Bank agreed to further strengthen their partnership to advance India's green-growth agenda at a meeting between Minister of State for Environment & Forests Jairam Ramesh, and World Bank Group President Robert B. Zoellick on 13.1.2011.

The partnership between the World Bank and the Government of India will support programs that help India maintain high growth as a primary objective, while also either building climate resilience or promoting clean energy solutions, consistent with India's 11th Five Year Plan. This includes projects that build fuel-efficient transport infrastructure, clean energy hydropower plants, efficient water supply and wastewater systems, as well as programs that help farmers, fishing communities and people in other vulnerable communities safeguard their livelihoods against the vagaries of a changing climate.

As India accelerates its response to the environment challenges that face its growing economy, the Bank too has stepped up its financial support for the government's various environment management and protection programs. Some of the initiatives include

1. Industrial Pollution Management Project

The Bank also provided $65 million to help rehabilitate select industrially polluted sites and to support the development of a national framework for the remediation of polluted and orphaned contaminated sites. An ongoing $20 million project for energy efficiency also supports the phasing out of ozone-depleting substances.

2. The National Ganga Project

The Bank is preparing a $ 1 billion project to support the government's ambitious program to clean and conserve the Ganga, which involves operationalisation of the National Ganga River Basin Authority and associated state-level institutions. It will finance sewage collection and treatment and municipal solid waste management to help reduce pollution in the river.

3. Strengthening Capacity of CPCB/SPCBs

The proposed project to strengthen the capacity of the Pollution Control Boards will help build their skills and infrastructure, ensure their financial sustainability and set up new monitoring and data management mechanisms.

Chapter 31

References

Abbate P.E., Dardanelli J.L., Cantarero M.G., Maturano M., Melchiori R.J.M. and Suero E.E. 2004. Climatic and water availability effects on water-use efficiency in wheat. *Crop Sci.*, **44:** 474–483.

Adhya T.K., Rath A.K., Gupta P.K., Rao V.R., Das S.N., Parida K.M., Parashar D.C. and Sethunathan N. (1994). Methane emission from flooded rice fields under irrigated conditions. *Biol. Fertil. Soils.* **18:** 245-248.

Ahmadi, A. and Baker D.A. 2001. The effect of water stress on grain filling processes in wheat. *Journal of Agricultural Science,* **136:** 257-269.

Allen, L.H. 1990. Plant responses to rising carbon dioxide and potential interactions with air pollutants. *J. Environmental Quality,* **19:** 15-34.

Antonellini M., Lee A. and Hotz S. 2008. Salt water intrusion in the coastal aquifer of the southern Po Plain, Italy. *Hydrogeol J.,* **16:** 1541–1556.

Athar H.R. and Ashraf A. 2009. Strategies for crop improvement against salinity and drought stress: An Overview. In: Athar HR, Ozturk M (eds): Salinity and water stress: improving crop efficiency. Springer, New York, pp 1–16.

Atwell B.J. and Steer B.T. 1990. The effect of oxygen deficiency on uptake and distribution of nutrients in maize plants. *Plant Soil,* **122:** p. 1–8.

Awmack C.S., Woodcock C.M. and Harrington R. 1997. Climate change may increase vulnerability of aphids to natural enemies. *Ecol Entomol.,* **22:** 366–368.

Barber S.A. 1995. Soil Nutrient Bioavailability: A Mechanistic Approach, 2nd Edn. John Wiley and Sons, Inc, New York, pp 414.

Barrios S., Ouattara B. and Strobl E. 2008. The impact of climatic change on agricultural production: Is it different for Africa? Food Policy, **33:** 287–298.

Bassirirad H. 2000. Kinetics of nutrient uptake by roots: responses to global change. *New Phytol.,* **147:** 155–169.

Birdsey R.A. 1996. Regional estimates of timber volume and forest carbon for fully stocked timberland, average management after final clear cut harvest. In Forests and Global Change: Volume 2, Forest Management Opportunities for Mitigating Carbon Emissions, eds. R.N. Sampson and D. Hair, American Forests, Washington, DC.

Black V.J., Black C.R., Roberts J.A. and Stewart C.A. 2000. Impact of ozone on the reproductive development of plants. *New Phytologist,* **147:** 421-447.

Booker F.L., Miller J.E. and Fiscus E.L. 1992. Effects of ozone and UV-B radiation on pigments, biomass and peroxidase activity in soybean. In: *Tropospheric Ozone and the Environment: II. Effects, Modelling and Control.* R.L. Berglund (ed.). Air and Waste Management Association, Pittsburgh, PA. pp. 489-503.

Bornman J.F. 1989. Target sites of UV-B radiation in photosynthesis of higher plants. *J. Photochem. Photobiol.,* **4:** 45-158.

Bose T.K., Mitra S.K., Farooq A.A. and Sadhu M.K. 1999. Tropical Fruits. Vol. 1. Nayaprakash, Bidhan Sarani, Calcutta, India.

Brouder S.M. and Volenec J.J. 2008. Impact of climate change on crop nutrient and water use efficiencies. *Physiological Plantarum* **133:** 705. 2008.

Cannell M.G.R., Grace J. and Booth A. 1989. Possible impacts of climatic warming on trees and forests in the United Kingdom: a review. *Forestry,* **62:** 337-364.

Chadha K.L., Singh A.K. and Patel V.B. 2011. Horticulture to Horti-Business. The Horticulture Society of India, Westville Publishing House. pp. 31-41.

Chapin F.S. 2003. Effects of plant traits on ecosystem and regional processes: a conceptual framework for predicting the consequences of global climate change. *Ann Bot.,* **91:** 455–463.

Cicerrone R.J. and Oremland R.S. 1988. Biogeochemical aspects of atmospheric methane. Global Biogeochem Cycles **2:** 299-327.

Cleland E.E., Chuine I., Menzel A., Mooney H.A. and Schwartz M.D. 2007. Shifting plant penology in response to global change. *Trends Ecol Evol.,* **22:** 357–365.

Cook A.C., Tissue D.T., Roberts S.W. and Oechel W.C. 1998. Effects of long-term elevated [CO_2] from natural CO_2 springs on Nardusstricta: photosynthesis, biochemistry, growth and penology. *Plant Cell Environ.*, **21:** 417–425.

Correia C., Torres–Pereira M. and Torres-Pereira M.S. 1999. Growth, photosynthesis and UV-B absorbing compounds of Portuguese barbela wheat exposed to UV-B radiation. *Environ. Pollut.*, **104:** 383– 388.

Cosgrove B. 1991. Weather: Dorling Kindersley Eyewitness Guides. Dorling Kindersley Ltd., Henrietta Street, London, 64p.

Costa H., Gallego T. 2002. Effects of UV-B radiation on antioxidant defense system in sunflower cotyledons. *Plant Sci.*, **162(6):** 939-945.

Coviella C., Trumble J. 1999. Effects of elevated atmospheric carbon dioxide on insect plant interactions. *Conserv Biol.*, **13:** 700–712.

Cramer M.D., Hawkins H.J. and Verboom G.A. 2009. The importance of nutritional regulation of plant water flux. Oecologia, **161:** 15–24.

Crutzen, P.J. 1995. On the role of methane in atmospheric chemistry: Sources, sinks and possible reduction in anthropogenic sources. *Ambio.*, **24:** 52-55.

Dar, D.W. 2007. Carbon sequestration for better dry land productivity. The Hindu In. 24.10.07 p.13.

Darwin R., Tsigas M., Lewandrowski J. and Raneses A. 1995. World Agriculture and Climate Change: Economic Adaptations. USDA Econ Res Serv, Washington, DC, AER-703 (June).

Das K., Panda K.D., Sarkar R.K. and Reddy J.N. 2009. Submergence tolerance in relation to variable floodwater conditions in rice. Environ. Exp. *Bot.*, **66:** 425–34.

Davidson E.A. and Janssens I.A. 2006. Temperature sensitivity of soil carbon decomposition and feedbacks to climate change. *Nature*, **440:** 165-173.

Drake B.G. and Leadley P.W. 1991. Canopy photosynthesis of crops and native plant communities exposed to long-term elevated CO_2 treatment. *Plant, Cell and Environment*, **14:** 853-860.

Eamus D. 1991. The interaction of rising CO_2 and temperatures with water-use efficiency. *Plant, Cell and Enviro.*, **14:** 843-852.

Eamus D. and Murray M. 1991. Photosynthetic and stomatal conductance responses of Norway Spruce and Beech to ozone acid mist and frost in a conceptual model. *Environmental Pollution*, 72: 23-44.

Easterling W. and Apps. M. 2005. Assessing the consequences of climate change for food and forest resources: A view from the IPCC. *Clim Change*, **70:** 165–189.

Egli E.B. 1998. *Seed Biology and Yield of Grain Crops*. New York: CAB International.

Evans L.T. 1975. The physiological basis of crop yield. In *Crop Physiology: Some Case Histories,* eds. L.T. Evans. Cambridge, UK: Cambridge University Press, pp. 327-355.

Fiscus E.L., Miller J.E. and Booker F.L.1994. Is UV-B a hazard to soybean photosynthesis and yield?: Results of a ozone/UV-B interaction study and model predictions. In: Stratospheric Ozone Depletion/ UV-B Radiation in the Biosphere. R.H. Biggs and M.E.B.Joyner (ed.). Springer-Verlag, Berlin. pp. 135-147.

Flynn D.F.B, Sudderth E.A. and Bazzaz F.A. 2006. Effects of aphid herbivory on biomass and leaf-level physiology of *Solanum dulcamara* under elevated temperature and CO_2. *Environ Exp Bot.,* **56:** 10–18.

Gonzalez R., Paul, Percy, Ambrose, McLaughlin, Barnes, Areses, Wellburn. 1996. Responses to ultraviolet-B radiation (289±315 nm) of pea (*Pisum sativum* L.) lines differing in leaf surface wax. *Physiologia Plantarum,* **98:** 852-860.

Gupta I.D. 2008. Causes and patterns of climate change. pp-39:97.

Heagle A.S., Heck W.W., Rawlings J.O. and Philbeck R.B. 1983. Effects of chronic doses of ozone and sulfur dioxide on injury and yield of soybeans in open-top field chambers. *Crop Sci.,* **23:** 1184-1191.

Hitz S. and Smith J. 2004. Estimating global impacts from climate change. Global Environmental Change, **14:** 201–218.

IMD (Indian Meteorological Department) 1957. Agricultural Meteorology in India Meteorological Department: A Review. Issued in Commemoration of Silver Jubilee of Agricultural Meteorology, Division of India Meteorological Department, 42p.

IPCC (Intergovernmental Panel on Climatic Change) 2006. The Economics of Climate Change: Stern Review. The Summary of conclusions. Survey of the Environment, 2007. The Hindu, pp 141-145.

IPCC, 2007. Working Group II. Fourth Assessment Report. Cambridge University Press, Cambridge.

IPCC, 1996. In Climate Change. The Science of Climate Change Report of Working Group I. (Houghton, J.J., Meira, L.G., Filho, B., Callander,

A., Harris, N., Jattenberg, A and Maskell, K. eds) Cambridge University Press, New York, pp.4.

IPCC, 2001. Climate Change: The scientific basis. Cambridge, UK, Cambridge University Press.

Izaurralde R.C., Rosenberg N.J., Brown R.A., Thomson A.M. 2003. Integrated assessment of Hadley Center (HadCM2) climate-change impacts on agricultural productivity and irrigation water supply in the conterminous United States. *Agric for Meteorol.*, **117:** 97–122.

Jan F., Feenstra, Ian Burton, Joel Smith B. and Richard S.J. Tol. 1998. Handbook on Methods for Climate Change Impact Assessment and Adaptation Strategies. pp: 230-257.

Johannsen W. 1903. Adaptation through natural selection. In: Crop adaptation and distribution. Eurasia Publishing House (P) Ltd, New Delhi, 448p.

Kabat, Pavel, Henk and Schaik, 2003. Climate changes the water rules: How water managers can cope with today's climate variability and tomorrow's climate change. Dialogue on Water and Climate: The Netherlands.

Karmakar R. 2016. Potential effects of climate change on soil properties: A Review Science International Volume 4 Issue 2, 2016.

Kennedy G., Nantel G. and Shetty P. 2003. The scourge of "hidden hunger": global dimensions of micronutrient deficiencies. Food, *Nutr Agric.*, **32:** 8–16.

Krupa S.V. and Kickert R.N. 1989. The greenhouse effect: impacts of UV-B radiation, CO_2 and O_3 on vegetation, *Environmental Pollution* **61:** 263-393.

Kuan meng goh 2004. Carbon sequestration and stabilization in soils: Implication for soil productivity and climate change. *Soil Sci. Plant Nutr.*, **50(4):** 467-476.

Kumar N., Singh S. and Dass A. 2009. Influence of Climate on Availability of Plant Nutrients in Soil. Indian Farmer's digest. pp 13-15.

Lal R. 2000. Soil management in the developing countries. *Soil Sci.*, **165:** 57–72.

Lal R. 2006. Enhancing crop yields in the developing countries through restoration of the soil organic carbon pool in agricultural lands. *Land Degrad Develop.*, **17:** 197–209.

Lal R. 1995. In Soil and Global Change (Lal, R., Kimble, J.M., Levine, E. and Stewart, B.A. eds), CRC Boca Raton, Florida, pp. 131-141.

Lal R. 2001. In Soil Carbon Sequestration and Green House Effect. SSSA Special Publication 57, Soil Science Society of America, Madison, pp 1-8.

Larson W.E. and Robert P.C. 1991. Farming by soil. In: Soil management for sustainability, (Lal, R and Pierce, F.J., Eds), *Soil Water Conserv. Soc., Ankeny, IA*. pp. 103-111

Lenka S., Lenka N.K., Kundhu, S. and Subba Rao A. 2013. Climate change and natural Resource Management. New India Publishing Agency, New Delhi. pp: 294-303.

Long S.P. 1991. Modification of the response of photosynthetic productivity to rising temperature by atmospheric CO_2 concentrations: has its importance been underestimated? *Plant, Cell and Environment,* **14:** 729-739.

Lynch J.P. and Brown K.M. 2001. Topsoil foraging - an architectural adaptation of plants to low phosphorus availability. *Plant Soil,* **237:** 225–237.

Lynch J.P. and St Clair S.B. 2004. Mineral stress: the missing link in understanding how global climate change will affect plants in real world soils. *Field Crops Res.,* **90:** 101–115

Mehra M., Kushwaha H.S. and Kumar A. 2010. Carbon sequestration in soil. *Indian Farmers digest.* 8-11.

Mepsted R., Paul, Stephen, Nogues, Corlett, Ayres, Jones, Baker. 1996. Effects of enhanced UV-B radiation on pea (*Pisum sativum* L.) grown under field conditions. *Global Change Biol.,* **2:** 325-334.

Mertz O., Halsnaes K., Olesen J.E. and Rasmussen K. 2009. Adaptation to Climate Change in Developing Countries. *Environ Manag.,* **43:** 743–752.

Mitchell R.A.C., Mitchell V.J., Driscoll S.P., Franklin J. and Lawlor D.W. 1993. Effects of increased CO_2 concentration and temperature on growth and yield of winter wheat at two levels of nitrogen application. *Plant Cell Environ.,* **16:** 521–529.

Mitra A.P. 1992. Greenhouse Gas Emission in India. Editor: A P Mitra. Global Change Scientific Rep. No.4, August, CSIR.

More, T.A. and Bhargava R. 2010. Impact of climate change on productivity of fruit crops in arid regions. In: Challenges of Climate Change-Indian Horticulture (Eds, Singh, H.P., Singh, J.P. and Lal, S.S) Westville Publishing House, New Delhi.

Morison J.I.L. 1985. Sensitivity of stomata and water-use efficiency to high CO_2. *Plant, Cell and Environ.,* **8:** 467-474.

Nord E.A. and Lynch J.P. 2008. Delayed reproduction in Arabidopsis thaliana improves fitness in soil with suboptimal phosphorus availability. *Plant Cell Environ.,* **31:** 1432–1441.

Nord E.A. and Lynch J.P. 2009. Plant phenology: a critical controller of soil resource acquisition. *J Exp. Bot.,* **60:** 1927–1937.

Patra D.D. 2007. Soil Carbon Sequestration in relation to soil and Environmental quality and food security. In. management of soil quality for sustainable Agriculture (eds. Misra, B., Chandra, R and Singh, S.K). Satish Serial Publishing House, New Delhi pp. 95-104.

Peter K.V. 2009. Basics of Horticulture. New India Publishing Agency. New Delhi. p. 47-70.

Pinjari S.S., Pawar P., Chavan, S.A and Talathi, M.J. 2007. Methane emission from paddy fields and its mitigating measures. *Agricultural update,* **2:** 17-18.

Prade K and Trolldenier G. 1990. Denitrification in the rhizosphere of rice and wheat seedlings as influenced by plant K status, air-filled porosity and substrate organic matter. *Soil Biol. Biochem.,* **22:** 769–773.

Prasad P.V.V., Boote K.J., Allen Jr. L.H. and Thomas J.M.G. 2002. Effects of elevated temperature and carbon dioxide on seed-set and yield of kidney bean (*Phaseolus vulgaris L.*). *Global Change Biology,* **8:** 710-721.

Prasad R. and Rana R. 2006. A study on maximum temperature during March 2004 and its impact on rabi crops in Himachal Pradesh. *J. of Agrometeorology,* **8(1):** 91-99.

Price P.W., Bouton C.E., Gross P., McPheron B.A., Thompson J.N. and Weis A.E. 1980. Interaction among three trophic levels: inuence of plants on interactions between insect herbivores and natural enemies. *Annu Rev Ecol Syst.,* **11:** 41–65.

Rao M.S., Srinivas K., Vanaja M., Rao G.G.S.N. and Venkateshwaralu B. 2008. Impact of elevated CO_2 on insect herbivores and host interactions, Research bulletin. CRIDA, Hyderabad, 36p.

Rao G.S.L.H.V.P. 2002. Climate variability and coconut production in the humid tropics. Kerala Agricultural University, Thrissur, India. 38p.

Rao, G.S.L.H.V.P. and Nair, R.R 1988. Agrometeorology of Plantation Crops. Kerala Agricultural University, Thrissur, Kerala, India. 187p.

Rayner N.A. 2003. Global analyses of sea surface temperature, sea ice, and night marine air temperature since the late nineteenth century. *J. Geophys. Res. Atmos.*,**108**: 440.

Rodhe H. 1990. A comparison of the contribution of various gases to the greenhouse effect. *Science*, **248:** 1217-1219.

Rosenzweig C. and Parry M.L. 1994. Potential impact of climate change on world food supply. *Nature*, **367:** 133–138.

Runeckles V.C. and Krupa, S.V. 1994. The impact of UV-B radiation and ozone on terrestrial vegetation. *Environmental Pollution*, **83:** 191-213.

Schimel J., Balser T.C. and Wallenstein M. 2007. Microbial stress response physiology and its implications for ecosystem function. *Ecology*, **88:** 1386–1394.

Sethunathan N., Neue H.E., Wassman R., Rao V.R., Adhya T.K., Ramakrishnan B., Jain M.C., Sushil Kumar, Murthy P.S.S. and Anthoni Raj S. 1996. Methane emission from rice fields. Paper presented in IRRI-India conference. New Delhi.

Setter T.L., Val R. and Mando A. 2009. Review of wheat improvement for waterlogging tolerance in Australia and India: the importance of anaerobiosis and element toxicities associated with different soils. *Ann Bot.*, **103:** 221–235.

Sharma J.S., Sing G. and Ramakrishna Y.S. 2004. Cold wave during 2003-03 over north Indian and its effect on crop. The Hindu, 10th January, 2004, p.6.

Shukla P.R., Sharma S.K. and Ramana V.P. 2002. Climate Change and India-Issues, Concerns and opportunities. Tata McGraw-Hill, New Delhi, p. 314.

Sing A.K. and Reddy N.N. 2010. Natural resource management and the ways to overcome abiotic stresses in fruit crops. Paper presented at National Seminar on Impact of Climate Change on fruit crops, PAU, Ludhiana. September, 4-6: 2010.

Singh B., El Maayar M., Andre P., Bryant C.R. and Thouez J.P. 1998. Impacts of a GHG-induced climate change on crop yields: Effects of acceleration in maturation, moisture stress and optimal temperature. *Clim Change*, **38:** 51–86.

Smith W.N., Grant B.B., Desjardins R.L., Qian B., Hutchinson J. and Gameda S. 2009. Potential impact of climate change on carbon in agricultural soils in Canada 2000–2099. *Clim Change*, **93:** 319–333.

Squire G.J. and Unsworth M.H. 1988. Effects of CO_2 and Climate Change on Agriculture. Report to U.K. Department of the Environment. University of Nottingham, England.

Sun B., Chen D.L., Li Y. and Wang X.X. 2008. Nitrogen leaching in an upland cropping system on an acid soil in subtropical China: lysimeter measurements and simulation. *Nutri. Cycl. Agroecosyst.*, **81:** 291–303.

Swamy M. and Bhattacharya S. 2011. Budgeting anthropogenic green house gas emission from Indian Livestock's using country-specific emission coefficients. *Current Sci.*, **91:** 1340-1353.

Temple P.J., Taylor O.C. and Benoit L.F. 1985. Cotton yield responses to ozone as mediated by soil moisture and evapotranspiration. J. *Environmental Quality,* **14:** 55-60.

Teramura A.H., Sullivan J.H. and Ziska L.H. 1990. Interaction of elevated ultraviolet-B radiation and CO_2 on productivity and photosynthetic characteristics in wheat, rice, and soybean. *Plant Physiology,* **94:** 470-475.

Turner A.G. and Annamalai H. 2012. Climate Change and the South Asian Monsoon, *Nature Climate Change.,* **2**: 587-595.

Uppala S.M. 2005. The ERA40 reanalysis, *Q.J.R. Meteorol. Soc.,* **131(612):** 2961–3012.

US EPA. 1995. *Air Quality Criteria for Ozone and other Photochemical Oxidants.* Draft edition. Research Triangle Park, North Carolina. Environmental Criteria and Assessment Office.

Van De Geijn S.C., Goudriaan J. and Berendse F. (eds.). 1993. *Climate Change; Crops* and *Terrestrial Ecosystems.* CABO-DLO, Wageningen. 144 p.

Vikas Abrol, Peeyush Sharma and Ravindra Kumar. 2012. Department of Soil Science and Agricultural Chemistry, Dryland Research Sub Station, India. Role of soil carbon in mitigating global house gas (GHG) emission. Rashtriya Krishi. **7:** 1-4.

Wardlaw I.F. 1990. The control of carbon partitioning in plants. *New Physiologist,* **116:** 341-381.

White P.J. and Broadley M.R. 2005. Bio-fortifying crops with essential mineral elements. *Trends Plant Sci.,* **10:** 586–593.

World Water Assessment Programme. 2003. Water for People, Water for Life: The United Nations World Water Development Report. UNESCO: Paris.

WRI (World Resource Institute). 1998. Mitigating climate change through forest and land use activities.

Yatagai A., Kamiguchi K., Arakawa O., Hamada A., Yasutomi N. and Kitoh A. 2012. APHRODITE: Constructing a Long-Term Daily Gridded Precipitation Dataset for Asia Based on a Dense Network of Rain Gauges. *Bull. Amer. Meteor. Soc.*, **93**: 1401-1415.

Zougmore R., Mando A. and Stroosnijder L. 2009. Soil Nutrient and Sediment Loss as Affected By Erosion Barriers and Nutrient Source in Semi-Arid Burkina Faso. *Arid Land Res Manag.*, **23:** 85–101.

❑❑❑

About the Authors

Dr. K. Balakrishnan is presently working as Professor of Crop Physiology at Agricultural College and Research Institute, Madurai. He was the recipient of ICAR Senior Fellowship during 1983 and obtained Ph.D in Crop Physiology in 1987 from Tamil Nadu Agricultural University, Coimbatore (TNAU). He has also obtained Post Graduate Diploma in Journalism in 1988. He also worked in United Planter's Association of South India, Coonoor, Nilgiris, Tamil Nadu. He has been teaching courses on Crop Physiology and Biochemistry to the Undergraduate and Postgraduate students in both Agriculture and Horticulture for the past 32 years. He served as Principle Investigator of Foliar diagnosis in Horticultural crops in plant scheme funded by Government of Tamil Nadu, Post Harvest Technology of Horticultural crops ICAR-PHT at Horticultural College and Research Institute, Periyakulam and Arid zone fruit crops funded by AICRIP (ICAR) , at Regional Agricultural Research Station, Aruppukkottai. He operated several schemes from both Private and Government agencies. He also worked as Assistant plant physiologist in ICAR-Tropical fruit crops, AICRIP, Horticultural College, Coimbatore. He was the member of All India ICAR Syllabus Committee on basic Agricultural sciences during 2007-2008. He also served as a member of Board of studies, Tamil Nadu Agricultural University during 1997-2000. He has authored and co-authored 10 book chapters and 6 books and published 60 research papers in both National and International journals, 3 extension bulletins and 120 popular articles. He has been the question setter and external examiner for M.Sc.(Ag.) and Ph.D thesis in several Universities. He has organized one national seminar at AC&RI, Madurai during September 2013. He served as the Professor and Head, Department of Seed Science and Technology, Agricultural College and Research Institute, Madurai during 2013-2015.

Professor Dr. M. Prakash born in 1966, obtained his B.Sc.(Ag.) and M.Sc.(Ag.) from Tamil Nadu Agricultural University, Coimbatore, India in 1988 and 1990 respectively. He did his Ph.D from Annamalai University. His studious and scholarly aptitude during his studenthood fetched him prestigious student support awards like National Merit Scholarship, Government of India during 1981-1988 and ICAR Junior Fellowship during 1988-1990. Prof. M. Prakash is working on abiotic stress tolerance from his post graduation onwards. He started working on effect of antitranspirants on growth and physiology of brinjal and continuing his work on abiotic stress tolerance till date. He is having 24 years of teaching and research experience in the Department of Genetics and Plant Breeding wherein he has successfully guided 8 Ph. D. and 22 Masters Degree Programmes. He has also completed five funded research projects.

He is presently serving as Professor, Department of Genetics and Plant Breeding, Faculty of Agriculture, Annamalai University. He has earned the status of Departmental Research Support with Special Assistance Programme of University Grants Commission (UGC-SAP-DRS) to the Department with financial assistance of Rs. 50 lakhs at Phase I (2009-2014) and Rs. 1.025 Crore at phase II (2015-2020) level.

His credentials include publications of 85 refereed research papers, seven books published by National publishers, three books published by International publishers and one book chapter published in SESAME by globally reputed CRC Press. He has significantly contributed in knowledge sharing, through organisation of 7 National Seminars, participation in 8 International Seminars, Conferences and Workshops and 18 National Seminars and Conferences. His expertise in abiotic stress tolerance is well recognised by invitations to serve in the Editorial Board/ Committees of several academic journals, awards like Crop Research Award in 2004 and Best Researcher Award in Annamalai University in 2010 and Fellowships like FELLOW of Indian Society of Plant Physiology, New Delhi in 2015 (FISPP), FELLOW of Association for the Advancement of Biodiversity Science, Bangalore in 2015 (FABSC) and FELLOW of National Academy of Biological Sciences (NABS), Chennai in 2016 (FNABS).

Dr.P.Jeyakumar is currently serving as Professor and Head, Department of Crop Physiology, Tamil Nadu Agricultural University, Coimbatore. He has been actively involved in teaching, research and extension activities for the last 22 years. He is specialized in the area of nutrio-hormonal and postharvest physiology. He has guided ten PG students and four Ph.D scholars. He served as Principal Investigator for more than 15 externally funded research projects and Co-Principal Investigator for 12 projects. He has undergone postdoctoral trainings in Postharvest Physiology at McGill University, Canada and ARO Volcani Centre, Israel. He has received the Best Researcher Award from Tamil Nadu Agricultural University, Coimbatore and Eminent Scientist Award from National Environmental Science Academy, New Delhi. He is also the recipient of Best Researcher Award in Agricultural Botany from MASU, TNAU, Coimbatore. He has authored and co-authored 15 book chapters and 7 books, and published more than 70 research papers in National and International journals. He has visited many international institutes in Australia, Canada, China, Israel, Italy, Malaysia, Turkey and Vietnam to present research papers and to participate in trainings. Prof.

P. Jeyakumar is serving as the Technical Editor of The Madras Agricultural Journal and serving as editorial committee member of many scientific journals. Currently he is working on postharvest physiology of tropical fruits and abioic stress management in vegetable crops. He has been the question setter and external examiner for M.Sc.(Ag.) and Ph.D thesis in several Universities. Co-ordinated the ISHS International Symposium on Papaya, Stress Physiology and three ICAR short courses: Nutrio-hormonal dynamics, Emerging techniques in fruit crops and Diagnosis and management of nutritional disorders. He has served as the Course Director for the ICAR Short course on Applied Crop Physiology and organized two National Seminars on Nutrio-physiology and Abiotic stress Management. He is also associated with two major projects on Department development, funded by DST- FIST and UGC- SAP, New Delhi.

□□□